NEUROMETHODS

Series Editor
Wolfgang Walz
University of Saskatchewan
Saskatoon, Canada

For further volumes:
http://www.springer.com/series/7657

Neuromethods publishes cutting-edge methods and protocols in all areas of neuroscience as well as translational neurological and mental research. Each volume in the series offers tested laboratory protocols, step-by-step methods for reproducible lab experiments and addresses methodological controversies and pitfalls in order to aid neuroscientists in experimentation. *Neuromethods* focuses on traditional and emerging topics with wide-ranging implications to brain function, such as electrophysiology, neuroimaging, behavioral analysis, genomics, neurodegeneration, translational research and clinical trials. *Neuromethods* provides investigators and trainees with highly useful compendiums of key strategies and approaches for successful research in animal and human brain function including translational "bench to bedside" approaches to mental and neurological diseases.

Molecular Imaging for Brain Diseases

Edited by

Daichi Sone

Department of Psychiatry, Jikei University School of Medicine, Tokyo, Japan

 Humana Press

Editor
Daichi Sone
Department of Psychiatry
Jikei University School of Medicine
Tokyo, Japan

ISSN 0893-2336   ISSN 1940-6045   (electronic)
Neuromethods
ISBN 978-1-0716-4493-5   ISBN 978-1-0716-4494-2   (eBook)
https://doi.org/10.1007/978-1-0716-4494-2

This Humana imprint is published by the registered company Springer Science+Business Media, LLC, part of Springer Nature.
The registered company address is: 1 New York Plaza, New York, NY 10004, U.S.A.

If disposing of this product, please recycle the paper.

Preface to the Series

Experimental life sciences have two basic foundations: concepts and tools. The Neuromethods series focuses on the tools and techniques unique to the investigation of the nervous system and excitable cells. It will not, however, shortchange the concept side of things as care has been taken to integrate these tools within the context of the concepts and questions under investigation. In this way, the series is unique in that it not only collects protocols but also includes theoretical background information and critiques which led to the methods and their development. Thus it gives the reader a better understanding of the origin of the techniques and their potential future development. The Neuromethods publishing program strikes a balance between recent and exciting developments like those concerning new animal models of disease, imaging, in vivo methods, and more established techniques, including, for example, immunocytochemistry and electrophysiological technologies. New trainees in neurosciences still need a sound footing in these older methods in order to apply a critical approach to their results.

Under the guidance of its founders, Alan Boulton and Glen Baker, the Neuromethods series has been a success since its first volume published through Humana Press in 1985. The series continues to flourish through many changes over the years. It is now published under the umbrella of Springer Protocols. While methods involving brain research have changed a lot since the series started, the publishing environment and technology have changed even more radically. Neuromethods has the distinct layout and style of the Springer Protocols program, designed specifically for readability and ease of reference in a laboratory setting.

The careful application of methods is potentially the most important step in the process of scientific inquiry. In the past, new methodologies led the way in developing new disciplines in the biological and medical sciences. For example, Physiology emerged out of Anatomy in the nineteenth century by harnessing new methods based on the newly discovered phenomenon of electricity. Nowadays, the relationships between disciplines and methods are more complex. Methods are now widely shared between disciplines and research areas. New developments in electronic publishing make it possible for scientists that encounter new methods to quickly find sources of information electronically. The design of individual volumes and chapters in this series takes this new access technology into account. Springer Protocols makes it possible to download single protocols separately. In addition, Springer makes its print-on-demand technology available globally. A print copy can therefore be acquired quickly and for a competitive price anywhere in the world.

Saskatoon, Canada *Wolfgang Walz*

Preface

The field of neuroimaging has undergone profound transformations in recent decades, allowing us to peer into the complexities of the human brain with ever-increasing clarity. Among these advances, molecular neuroimaging stands at the forefront, bridging the gap between cellular-level biology and whole-brain function. By enabling the visualization of specific structures and function of the brain in vivo, this discipline offers unprecedented insight into the mechanisms underlying neurological and psychiatric disorders, as well as normal brain function.

This book "Molecular Imaging for Brain Diseases" aims to update this topic by providing timely reviews with special focuses on clinical aspects. Nuclear imaging, such as positron emission tomography (PET) or single-photon emission computerized tomography (SPECT), can target specific brain tissues using designated radioactive tracers. Proton magnetic resonance spectroscopy (1H-MRS) can be used to measure brain metabolite levels, e.g., N-acetyl aspartate (NAA), choline (Cho). The chapters reflect the interdisciplinary nature of the field, spanning chemistry, neuroscience, and clinical applications, and are authored by leading experts who are pushing the boundaries of what we know about the brain. The first half of the book (Chaps. 1, 2, 3, 4, 5, 6, 7, and 8) deals with advanced nuclear neuroimaging. It begins with chapters on glutamate receptor imaging (Chaps. 1 and 2), followed by tau and amyloid imaging (Chaps. 3, 4, and 5); in Chaps. 6, 7, and 8, glucose metabolism, opioid receptor, and dopamine receptor imaging are covered. The second half of the book, Chaps. 9, 10, and 11, describes methods using brain metabolite measurements by 1H-MRS. The final chapter, Chap. 12, is devoted to analytical methods using artificial intelligence, which have become extremely important in recent years.

As molecular neuroimaging continues to evolve, it holds promise for revolutionizing diagnostics, therapeutic monitoring, and personalized medicine. It is my hope that this book will not only provide readers with the technical knowledge needed to navigate this complex field but also inspire a deeper appreciation for the brain's molecular underpinnings and the technologies that make their study possible.

Tokyo, Japan   *Daichi Sone*

Contents

Contributors

SACHIKO ANAMIZU • *Department of Neuropsychiatry, Keio University School of Medicine, Tokyo, Japan; National Hospital Organization (NHO) Tochigi Medical Center, Tochigi, Japan*

IMAN BEHESHTI • *PrairieNeuro Research Centre, Kleysen Institute for Advanced Medicine, Health Sciences Centre, Winnipeg, MB, Canada; Department of Human Anatomy and Cell Science, Rady Faculty of Health Sciences, University of Manitoba, Winnipeg, MB, Canada*

HIRONOBU ENDO • *Advanced Neuroimaging Center (ANC), Institute for Quantum Medical Science, National Institutes for Quantum Science and Technology (QST), Chiba, Japan*

KJELL ERLANDSSON • *Institute of Nuclear Medicine, University College London, London, UK; Department of Neurology, Clinical Neuroscience Center, University Hospital and University of Zurich, Zurich, Switzerland*

MARIAN GALOVIC • *Institute of Nuclear Medicine, University College London, London, UK; Department of Neurology, Clinical Neuroscience Center, University Hospital and University of Zurich, Zurich, Switzerland*

JURI GELOVANI • *Faculty of Medicine Siriraj Hospital, Department of Radiology, Mahidol University, Bangkok, Thailand*

ARIEL GRAFF-GUERRERO • *Institute of Medical Science, University of Toronto, Toronto, ON, Canada*

MAKOTO HIGUCHI • *Advanced Neuroimaging Center (ANC), Institute for Quantum Medical Science, National Institutes for Quantum Science and Technology (QST), Chiba, Japan*

SIHONG HUANG • *Department of Radiology, The Second Xiangya Hospital, Central South University, Changsha, Hunan, China*

MASANORI ICHIHASHI • *Advanced Neuroimaging Center (ANC), Institute for Quantum Medical Science, National Institutes for Quantum Science and Technology (QST), Chiba, Japan*

YUKO KATAOKA • *Advanced Neuroimaging Center (ANC), Institute for Quantum Medical Science, National Institutes for Quantum Science and Technology (QST), Chiba, Japan*

FUMITOSHI KODAKA • *Department of Psychiatry, The Jikei University School of Medicine, Tokyo, Japan*

MATTHIAS KOEPP • *Department of Clinical and Experimental Epilepsy, UCL Queen Square Institute of Neurology, London, UK*

TERUKI KOIZUMI • *Institute of Medical Science, University of Toronto, Toronto, ON, Canada*

YUKI KOMATSU • *Advanced Neuroimaging Center (ANC), Institute for Quantum Medical Science, National Institutes for Quantum Science and Technology (QST), Chiba, Japan; Department of Neuropsychiatry, Keio University School of Medicine, Tokyo, Japan*

SHIN KUROSE • *Advanced Neuroimaging Center (ANC), Institute for Quantum Medical Science, National Institutes for Quantum Science and Technology (QST), Chiba, Japan; Department of Neuropsychiatry, Keio University School of Medicine, Tokyo, Japan*

JUN LIU • *Department of Radiology, The Second Xiangya Hospital, Central South University, Changsha, Hunan, China; Department of Radiology Quality Control*

Center, Hunan Province, Changsha, Hunan, China; Clinical Research Center for Medical Imaging in Hunan Province, Changsha, Hunan, China

HIROSHI MATSUDA • Department of Biofunctional Imaging, Fukushima Medical University, Fukushima, Japan

YUKI MOMOTA • Advanced Neuroimaging Center (ANC), Institute for Quantum Medical Science, National Institutes for Quantum Science and Technology (QST), Chiba, Japan; Department of Neuropsychiatry, Keio University School of Medicine, Tokyo, Japan

SHO MORIGUCHI • Advanced Neuroimaging Center (ANC), Institute for Quantum Medical Science, National Institutes for Quantum Science and Technology (QST), Chiba, Japan; Department of Neuropsychiatry, Keio University School of Medicine, Tokyo, Japan

WEERASAK MUANGPAISAN • Faculty of Medicine Siriraj Hospital, Department of Radiology, Mahidol University, Bangkok, Thailand

RODOLPHE NENERT • Departments of Neurology, University of Alabama at Birmingham (UAB) Heersink School of Medicine, Birmingham, AL, USA

JARRAD PERRON • Graduate Program in Biomedical Engineering, Price Faculty of Engineering, University of Manitoba, Winnipeg, MB, Canada; PrairieNeuro Research Centre, Kleysen Institute for Advanced Medicine, Health Sciences Centre, Winnipeg, MB, Canada

SIRIWAN PIYAPITTAYANAN • Faculty of Medicine Siriraj Hospital, Department of Radiology, Mahidol University, Bangkok, Thailand

NARUHIKO SAHARA • Advanced Neuroimaging Center (ANC), Institute for Quantum Medical Science, National Institutes for Quantum Science and Technology (QST), Chiba, Japan

CHAKMEEDAJ SETHANANDHA • Faculty of Medicine Siriraj Hospital, Department of Radiology, Mahidol University, Bangkok, Thailand

AYUSHE A. SHARMA • Department of Neurology, Yale University, School of Medicine, New Haven, CT, USA

TOSSAPORN SIRIPRAPA • Faculty of Medicine Siriraj Hospital, Department of Radiology, Mahidol University, Bangkok, Thailand

DAICHI SONE • Department of Psychiatry, Jikei University School of Medicine, Tokyo, Japan; Department of Clinical and Experimental Epilepsy, UCL Queen Square Institute of Neurology, London, UK

HISAOMI SUZUKI • Department of Neuropsychiatry, Keio University School of Medicine, Tokyo, Japan; National Hospital Organization (NHO) Shimofusa Psychiatric Medical Center, Chiba, Japan

JERZY P. SZAFLARSKI • Departments of Neurology, University of Alabama at Birmingham (UAB) Heersink School of Medicine, Birmingham, AL, USA; Departments of Neurobiology, UAB Epilepsy Center (UABEC), Birmingham, AL, USA; Departments of Neurology, UAB Epilepsy Center (UABEC), Birmingham, AL, USA

KENJI TAGAI • Advanced Neuroimaging Center (ANC), Institute for Quantum Medical Science, National Institutes for Quantum Science and Technology (QST), Chiba, Japan

TAKUYA TAKAHASHI • Department of Physiology, Yokohama City University Graduate School of Medicine, Yokohama, Japan

KEISUKE TAKAHATA • Advanced Neuroimaging Center (ANC), Institute for Quantum Medical Science, National Institutes for Quantum Science and Technology (QST), Chiba, Japan; Department of Neuropsychiatry, Keio University School of Medicine, Tokyo, Japan

TANYALUCK THIENTUNYAKIT • Faculty of Medicine Siriraj Hospital, Department of Radiology, Mahidol University, Bangkok, Thailand

THONNAPONG THONGPRAPARN • *Faculty of Medicine Siriraj Hospital, Department of Radiology, Mahidol University, Bangkok, Thailand*

NANTIKA WANNASOUPOL • *Faculty of Medicine Siriraj Hospital, Department of Radiology, Mahidol University, Bangkok, Thailand*

TENSHO YAMAO • *Department of Radiological Sciences, School of Health Sciences, Fukushima Medical University, Fukushima, Japan*

Chapter 1

Recent Progress in NMDA Glutamate Receptor Imaging

Kjell Erlandsson and Marian Galovic

Abstract

N-methyl-D-aspartate receptors (NMDARs) are involved in higher cognitive function of the brain, but also in the disease mechanisms of several neurological disorders. Over the past three decades, several groups have worked on the development of tracers for imaging NMDARs with single photon emission computed tomography (SPECT) or positron emission tomography (PET). These tracers only bind to open receptors, and thereby reflect both receptor density and activation status. It is not always straightforward to determine the level of specificity in the binding of an NMDAR tracer, and controversial findings have been reported. Quantification of tracer binding requires dynamic imaging and kinetic analysis. As NMDARs are ubiquitous in the brain, there is no reference region, and absolute quantification, based on a metabolite-corrected arterial input function, is needed. Despite a series of difficulties, multiple clinical conditions have been investigated using both SPECT and PET NMDAR tracers.

Key words N-methyl-D-aspartate receptor, Single photon emission computed tomography, Positron emission tomography, Kinetic analysis, Absolute quantification

1 Introduction

1.1 The NMDA Receptor

The N-methyl-D-aspartate receptor (NMDAR) is part of the glutamatergic system, the main excitatory neurotransmitter system in the brain, and is involved in fast synaptic transmission, neuronal plasticity and higher cognitive function, and plays a role in the disease mechanisms of several neurologic and psychiatric disorders [1, 2].

The NMDAR is an ion channel, which can be open or closed, i.e., active or inactive. In the inactive state, the channel is blocked by an Mg^{2+} ion bound to an intra-channel binding site. Activation occurs by removal of the Mg^{2+} ion, which requires simultaneous binding of the agonist glutamate and the co-agonist glycine as well as depolarization of the membrane potential (see Fig. 1) [3]. The opening of the receptor channel leads to the influx of Na^+ and Ca^{2+} ions and the efflux of K^+ ions. Overactivation leads to intracellular accumulation of Ca^{2+} ions, which can trigger a cascade of events

Daichi Sone (ed.), *Molecular Imaging for Brain Diseases*, Neuromethods, vol. 222, https://doi.org/10.1007/978-1-0716-4494-2_1,
© The Author(s), under exclusive license to Springer Science+Business Media, LLC, part of Springer Nature 2025

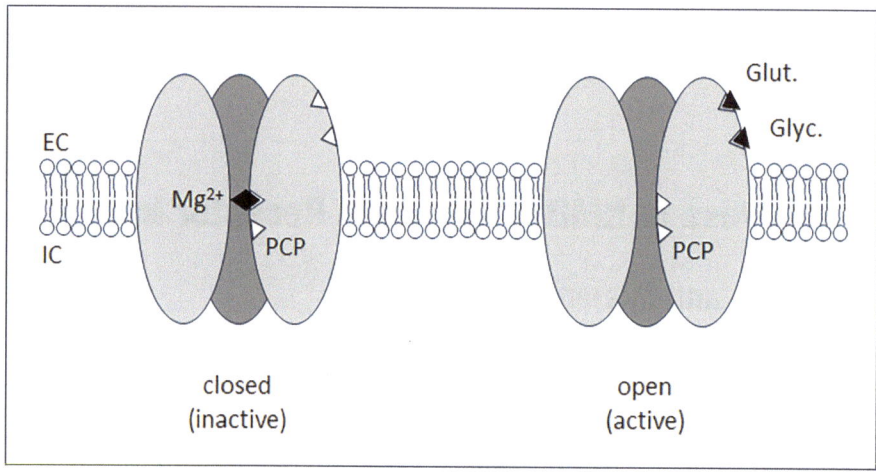

Fig. 1 Schematic illustration of an N-methyl-D-aspartate receptors (NMDAR) in the closed (left) and open state (right). The intra-channel phencyclidine (PCP)-binding site is also shown. The channel is closed by binding of Mg^{2+} and opened by binding of glutamate and glycine and depolarization of the membrane potential. The extra- (EC) and intracellular (IC) sides are as indicated

leading to cell death [4, 5]. The NMDA ion channel therefore plays an important role in both neuroprotection and neurodegeneration.

The NMDAR complex consists of several subunits, the composition of which is regionally variable. The receptor localization can be both synaptic and extra-synaptic, which may have some influence on their function. It has been suggested that activation of synaptic receptors is neuroprotective, whereas activation of extra-synaptic receptors may promote cell death, although conflicting findings have also been reported [6, 7].

1.2 Clinical Aspects   NMDARs are involved in the encoding of long-term memory through synaptic modification, and also in short-lasting working memory (WM), through maintained neuronal firing [8, 9]. Dysfunction of NMDARs has been proposed as hypotheses for the pathophysiology of both attention deficit hyperactivity disorder (ADHD) and schizophrenia [10, 11]. ADHD may be linked to the effect on WM (during neurodevelopment), which may also be the cause of age-related decline in memory performance [12]. Schizophrenia is believed to be related to NMDAR hypofunction [13], while anti-NMDA-receptor encephalitis is believed to be caused by receptor internalisation, leading to schizophrenia-like symptoms [14–16].

According to the widely accepted amyloid hypothesis of Alzheimer's disease (AD), the age-related neurodegenerative disorder is associated with increased accumulation of amyloid beta-peptide (Aβ), and it has been proposed that this could lead to the over-activation of extra-synaptic NMDARs, which would lead to influx of Ca^{2+} and subsequent synaptic loss [17, 18]. NMDARs have also been associated with the pathomechanisms of Parkinson's disease and Huntington's disease [19, 20].

NMDA antagonists have been used in the treatment of epilepsy but may, on the other hand, also have proconvulsant effects [21–24]. NMDA antagonists can also slow the growth of glutamate-secreting brain tumors [25].

1.3 Imaging NMDA Receptors

For many years, it has been of great interest within the research community to develop NMDAR tracers for the molecular imaging modalities single photon emission computed tomography (SPECT) and positron emission tomography (PET). A number of ligands have been developed that target an intra-channel binding site, associated with the noncompetitive antagonists phencyclidine (PCP), ketamine, and dizocilpine (MK-801). Binding to this site is only possible when the channel is open [13, 26]. Therefore, the binding of these tracers reflects not only receptor density but also activation status [27]. Far fewer NMDARs are open at rest than during activation. The PCP site is present on all types of NMDARs. Thus, tracer binding is not specific to any receptor subunit composition, and the tracers cannot distinguish between synaptic and extra-synaptic receptors. PET or SPECT can provide insights into the in vivo function of NMDARs in health and disease, assist in the development of novel treatment strategies, and can also be used to assess the effects of medication on NMDARs. We will refer to the intra-channel binding site as $NMDAR_{IC}$.

2 Materials

Over the years, a number of $NMDAR_{IC}$ SPECT and PET tracers have been developed. Here we review the ones which have been successful enough to be tested in humans. A common problem in the development of these tracers is demonstrating that they exhibit specific binding to $NMDAR_{IC}$. The search for new and better tracers is still ongoing. In Fig. 2, a schematic summary of various NMDAR SPECT and PET tracers is presented (see **Note 1**).

2.1 MK-801

One of the earliest compounds used is MK-801, which is an NMDAR antagonist with anticonvulsant and neuroprotective properties [28]. It has been labeled with both ^{123}I for SPECT studies and ^{18}F for PET [28–31]. It was tested in patients with cerebral ischemia and also in AD patients [29, 30]. It was found to be of limited utility due to high nonspecific binding due to high lipophilicity, and its binding to $NMDAR_{IC}$ was deemed inconclusive (see **Note 2**).

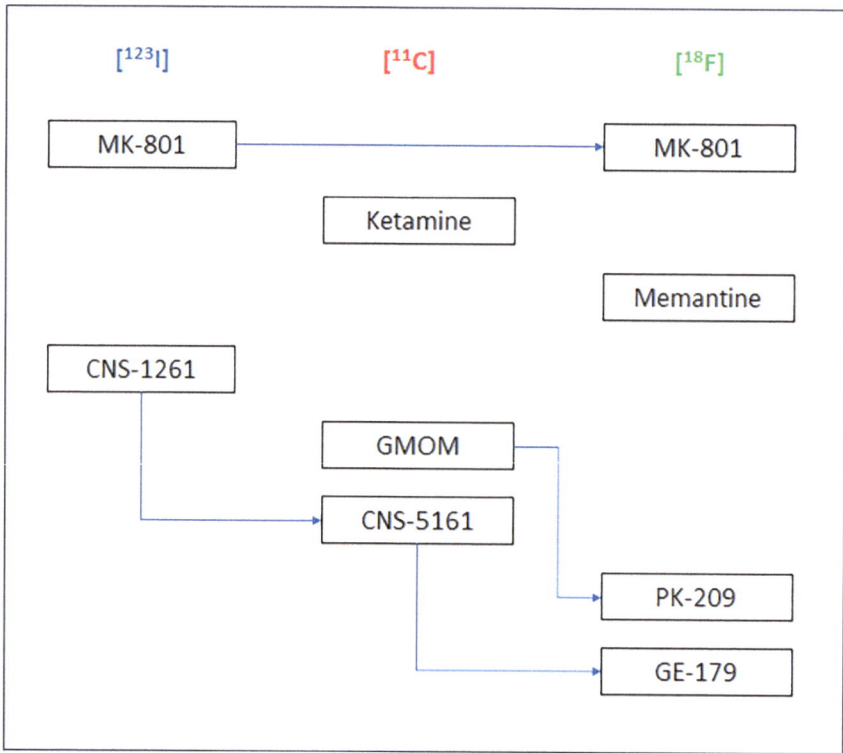

Fig. 2 Overview of various NMDAR single photon emission computed tomography (SPECT) and positron emission tomography (PET) tracers, labeled with [123]I (left), [11]C (center), or [18]F (right column). The vertical position indicates the order in which they were developed (from top to bottom)

2.2 Ketamine

Ketamine was labeled with ^{11}C to be used as a PET tracer. Specific binding to $NMDAR_{IC}$ was demonstrated in nonhuman primate (NHP) studies by displacement following pretreatment with unlabelled MK-801 [32]. The tracer was also used in studies on patients with temporal lobe epilepsy. A reduced tracer uptake was observed in the epileptic temporal lobe compared to the contralateral side. However, it was unclear whether this effect was due to a reduction in specific binding, neuronal loss, or reduced blood flow [33] (*see* **Note 3**).

2.3 Memantine

The drug memantine, used in the treatment of PD and movement disorders, was labeled with ^{18}F for PET studies. Decreased binding was found in NHPs after pretreatment with MK-801 [34]. Human studies were then performed in healthy volunteers (HV) [35]. The results were inconsistent, with a large inter-individual variation. It was concluded that the tracer uptake was highly perfusion-dependent and that its distribution in the brain did not reflect NMDAR concentration [35].

2.4 CNS-1261

The SPECT tracer CNS-1261, developed from the neuroprotective compound CNS-1102, was initially labeled with ^{125}I for preclinical studies. The uptake in the ischemic rat brain was similar to that of MK-801, while the lipophilicity was lower [36]. Human SPECT studies were performed in HVs with [^{123}I]CNS-1261, and the kinetic behavior of the tracer was characterized [37, 38]. Specific binding was demonstrated by displacement of the tracer with unlabeled ketamine in HVs. A significant reduction was found in the tracer binding metric "total volume of distribution" (V_T) in most of the brain regions examined ($p < 0.05$) [39]. As the effects of ketamine administration in humans are similar to both positive and negative psychotic symptoms of schizophrenia, possibly through the action of the drug on NMDAR, the subjects in the study were assessed with the Brief Psychiatric Rating Scale (BPRS). As a result, significant correlations were found between the negative BPRS subscale and reduction in V_T ($p < 0.01$) and also with a reduction in the more specific relative binding indices BI_1 and BI_2 ($p < 0.001$) [40] (*see* **Note 4**).

[^{123}I]CNS-1261 was also used to study schizophrenic patients, both drug-free and drug-treated [41]. A significant reduction in V_T was observed in clozapine-treated patients compared to healthy controls ($p < 0.005$). In patients treated with typical antipsychotics, there was also a reduction in V_T, but this was not significant. In drug-free patients, there was no apparent difference in V_T relative to healthy controls. However, using the outcome measures BI_1 and BI_2, which measure the specific tracer binding relative to the brain cortex, a significant reduction was found in the left hippocampus compared to controls ($p < 0.001$) [42].

2.5 GMOM

The PET tracer [^{11}C]GMOM was developed, and preclinical studies were performed [43]. Blocking studies in conscious rats by pre-injection of unlabeled GMOM and MK-801 showed a reduction in tracer binding ($p < 0.05$). However, no reduction was found in anesthetized NHPs after pre-administration of MK-801, although this could have been due to confounding effects of the anesthesia. [^{11}C]GMOM was later used in HVs, and a significant reduction in the net influx rate (K_i) after administration of ketamine was reported [44].

2.6 PK-209

An ^{18}F-labeled analog of GMOM, [^{18}F]PK-209, was developed, and PET studies in NHPs showed reduced V_T compared with baseline after pretreatment with MK-801 in two of three subjects [45]. When the tracer was tested in HVs, a relatively large test-retest variability in K_i was found. Furthermore, there was no consistent effect of ketamine administration on tracer binding [46] (*see* **Note 5**).

2.7 CNS-5161

$[^{11}C]$CNS-5161 was developed and evaluated in preclinical studies and in HVs [47, 48]. A significant correlation was found between V_T in the left hippocampus and verbal memory performance [49]. The tracer was also tested in PD patients. In patients taking the drug levodopa, there was a difference in tracer binding between those with and without dyskinesia, while in patients withdrawn from levodopa, no difference was observed [19].

2.8 GE-179

An ^{18}F-labeled analog of CNS-5161, $[^{18}F]$GE-179, was developed and characterized in HVs [50]. It was then tested in epileptic patients [51]. When compared to control subjects, there was a globally increased V_T in patients not taking antidepressants ($p < 0.002$), and globally reduced V_T in patients taking antidepressant drugs ($p < 0.002$). These findings are consistent with the hypothesis of overactive NMDAR in epilepsy and with a possible mode of action of the administered drugs. However, a possible confounding effect of nonspecific binding could not be excluded.

In a study performed in order to validate the specificity of $[^{18}F]$ GE-179 binding, unilateral electrical stimulation of the hippocampus in rats resulted in significantly increased tracer uptake compared to the contralateral hippocampus and to a control group of animals. Furthermore, this increase was blocked by the pre-injection of ketamine [52].

$[^{18}F]$GE-179 was also used in a study on patients with first-episode psychosis [53]. Hippocampal relative V_T was found to be significantly lower in the patients taking antipsychotics compared to healthy control subjects ($p = 0.006$). There was also a significant negative correlation between hippocampal relative V_T and PANSS general symptom severity ($p < 0.001$), but not with positive or negative symptom severity.

Finally, a recent study found a regional reduction in $[^{18}F]$GE-179 binding in patients recovering from NMDAR antibody encephalitis, while one recovered patient had slightly elevated binding, suggesting a rebound of NMDAR function [54].

Figure 3 shows the tracer uptake distribution in GE-179 studies of HV subjects. The uptake is ubiquitous throughout the brain, with higher values in subcortical structures as compared to the cortex.

3 Methods

3.1 Kinetic Analysis

After the administration of radiotracer, which is usually done intravenously, the uptake in different organs and tissues is determined by the delivery, retention, and clearance of the tracer. The rates of delivery to and clearance from tissue are affected by the blood flow. The retention is determined by specific or nonspecific binding or by the metabolism of tracer molecules.

SPECT and PET are two nuclear medicine imaging modalities that can be used to map the 3D distribution of a radiotracer in vivo in humans [55]. Tomographic images are generated using either analytical (e.g., filtered back-projection (FBP)) or iterative reconstruction algorithms (e.g., ordered subsets expectation maximization (OSEM)). It is important to apply corrections for photon attenuation and scatter, random coincidences (in PET), dead time, as well as motion and partial volume effects (PVE). The tracer uptake can be quantified by the standardized uptake value (SUV), corresponding to the radiotracer concentration (Bq/mL) in a volume of interest (VOI) or a volume element (voxel) in the image divided by the injected activity per unit of body weight of the subject (Bq/g). In normal clinical PET or SPECT scans, a single scan is performed sometime after the tracer injection to allow for tracer uptake in the tissue of interest. In neurological studies, the SUV will be dependent on blood flow, specific and nonspecific binding, as well as the uptake time. In order to obtain more specific quantitative parameters, dynamic studies can be performed.

The time course of the tracer concentration in tissue can be determined with dynamic PET or SPECT studies. This corresponds to a series of scans performed over a period of time, starting at the time of injection of the radiotracer. After quantitative reconstruction of tomographic images, time-activity curves (TACs) can then be generated either for different VOIs or for single voxels.

3.2 Compartmental Modeling

Kinetic analysis corresponds to analyzing the shape of the TACs, using an appropriate mathematical model, to determine quantitative values for various physiological or biochemical parameters. It is assumed that physiological processes affecting the measurement are in a steady state during the experiment and also that the radioligands are administered in tracer concentrations and therefore do not affect the biochemical processes being studied.

The mathematical models used in kinetic analysis are usually compartmental models [56]. Model-free techniques, also known as "graphical analysis," also exist. These are typically faster but provide less information and can be less accurate. In any case, it is necessary to have an input function, which ideally corresponds to the time course of tracer concentration in arterial plasma, known as the arterial input function (AIF).

When a compartmental model is used to describe a biological system, different compartments can represent either tracer contained in different spaces (e.g., intra-vascular, interstitial, or intracellular space) or, alternatively, tracer in different chemical states (e.g., free tracer in tissue or tracer molecules bound to specific or nonspecific binding sites). It is assumed that the tracer concentration within a compartment is homogeneous at all times. The tracer is transferred between compartments at a rate

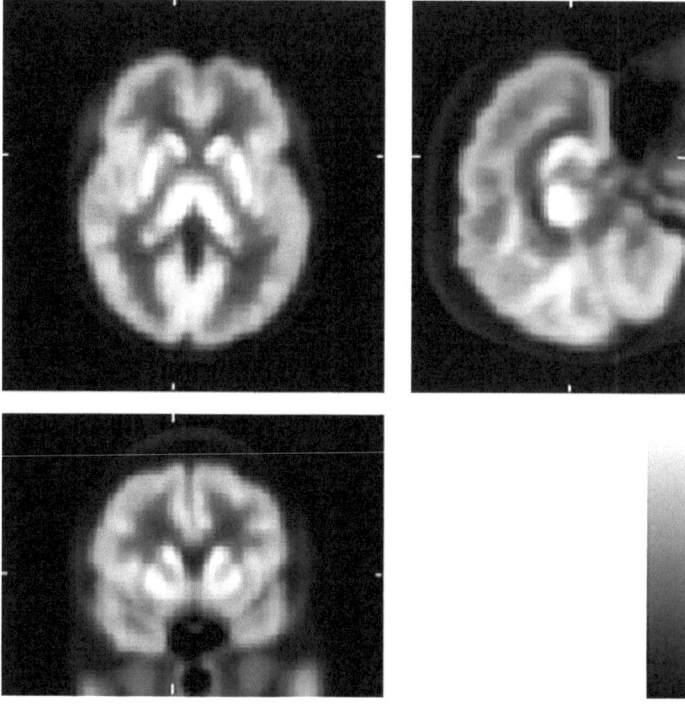

Fig. 3 Tracer uptake distribution in GE-179 studies of healthy volunteers (HVs); an average of ten subjects. Transaxial (top left), sagittal (top right), and coronal sections (bottom) are shown

proportional to the concentration in the compartment of origin and to a constant known as the rate constant. A compartmental system will therefore be defined by a set of differential equations.

Each model is referred to by the number of compartments representing tracers in tissue, or tissue compartments (TCs), as blood is not treated as a compartment from a mathematical point of view. You also distinguish between reversible and irreversible models, the latter being one in which there is a compartment with an input but no output.

A complete model for neuroreceptor studies would correspond to a reversible 3-TC model, including compartments for free tracer in tissue (F), specifically bound (S), and nonspecifically bound (NS) tracer (Fig. 3a). In practice, however, a simpler 2-TC model is normally used, where the NS compartment is "internalized" into the F compartment, which thereby becomes the non-displaceable (ND) compartment (Fig. 3b). Nonspecific binding typically corresponds to protein binding, which is non-saturable and non-displaceable. Specific binding, on the other hand, can be displaced by the un-labeled compound or by a drug that competes for the same binding sites. The 3-TC model has six rate constants (K_1, k_2, k_3, k_4, k_5, k_6), while the 2-TC model has four ($K_1, k_2{'}, k_3{'}, k_4$). Here we use primes to distinguish rate constants in the 2-TC model

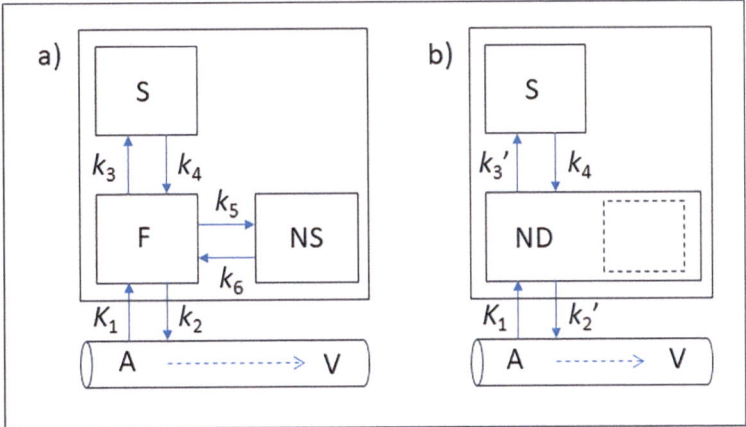

Fig. 4 Compartmental structure of the 3-TC (left) and 2-TC model (right), with 6 and 4 rate constants, respectively. Arterial (A) and venous (V) blood are also indicated

from those in the 3-TC model, a notation introduced by Koeppe et al.; however, this is not common practice [57]. Figure 4 shows the 3-TC and 2-TC models, including the various parameters.

The 2-TC model is described by the following differential equations:

$$
\begin{cases}
\frac{d}{dt} C_{\mathrm{ND}}(t) = K_1 C_p(t) - k_2' C_{\mathrm{ND}}(t) - k_3' C_{\mathrm{ND}}(t) + k_4 C_s(t) \\
\frac{d}{dt} C_S(t) = k_3' C_{\mathrm{ND}}(t) - k_4 C_s(t)
\end{cases}
\tag{1}
$$

where $C_{\mathrm{ND}}(t)$ and $C_S(t)$ are the tracer concentrations at time t in the ND and S compartments, respectively, and $C_p(t)$ is the AIF.

The rate constant K_1 is defined as the product of blood flow (F) and extraction fraction (E), which is dependent on blood flow and on the permeability-surface area product, while k_2 is defined as the ratio of K_1 and the fractional distribution volume of the F compartment (V_D):

$$
K_1 = F \cdot E; \quad k_2 = K_1 / V_D
\tag{2}
$$

It is important to note that both K_1 and k_2 depend on blood flow, but their ratio does not. K_1 is specified in the same units as blood flow (perfusion), e.g., [mL/min/mL], and all other rate constants in inverse units of time, e.g., [min^{-1}].

The rate constants k_3 and k_4 are defined based on the association and dissociation rate constants (k_{on} and k_{off}) from in vitro binding assay theory [58]. We therefore have $k_3 = k_{\mathrm{on}} \cdot B_{\mathrm{avail}}$ and $k_4 = k_{\mathrm{off}}$, where B_{avail} is the concentration of available binding sites (receptors), which can be lower than the total receptor concentration (B_{max}) due to the binding of an endogenous ligands or an administered drug. In the case of NMDAR$_{\mathrm{IC}}$ imaging, B_{avail} is the concentration of active NMDARs. In the 2-TC model, we have $k_i' = f_{ND} \cdot k_i$ for $i = 2, 3$, where f_{ND} is the free fraction in the ND compartment at equilibrium ($f_{ND} = [C_F / C_{ND}]_{eq}$).

By solving the system of differential equations, the total concentration of tracer in tissue, $C_T(t)$, is given by the convolution of the AIF with an impulse response function, $R(t)$, which can be expressed as the sum of mono-exponential functions, one for each compartment:

$$C_T(t) = C_p(t) \otimes R(t) = C_p(t) \otimes \left(\varphi_1 e^{-\theta_1 t} + \varphi_2 e^{-\theta_2 t} \right) \quad (3)$$

where φ_i and θ_i are expressions containing the rate constants.

In practice, it is necessary to take into account the presence of a certain amount of blood within the tissue, determined by the fractional blood volume, V_B [mL/mL]. The measured tissue concentration then becomes:

$$C_T(t) = (1 - V_B) \cdot C_p(t) \otimes R(t) + V_B \cdot C_B(t) \quad (4)$$

where $C_B(t)$ is the total activity concentration in blood.

The values of the rate constants are estimated by fitting the model equation to the measured TACs by minimizing a weighted

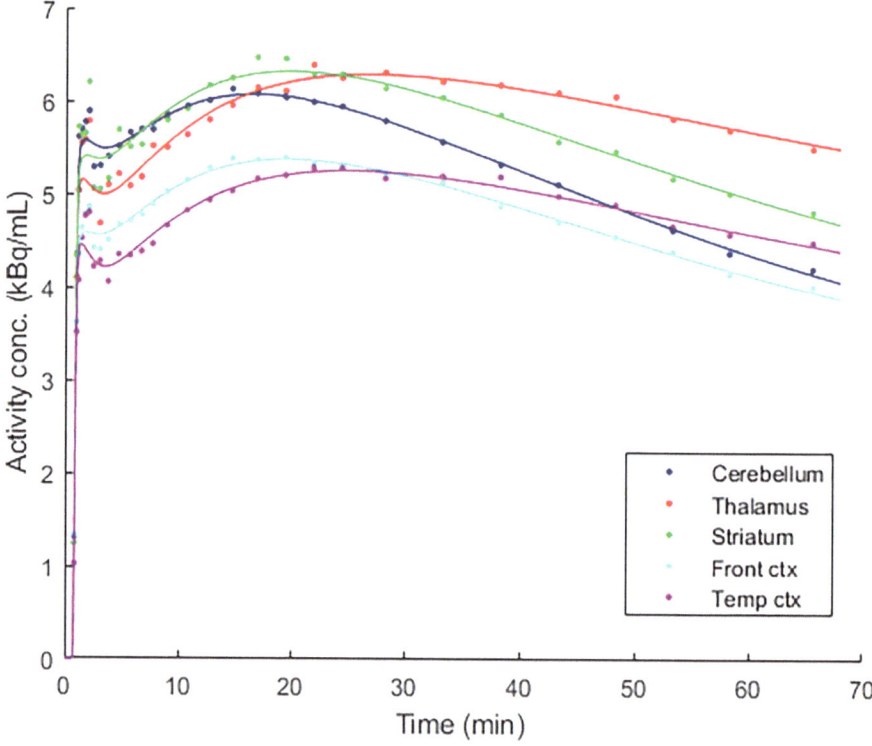

Fig. 5 Time-activity curves (TACs) from a GE-179 study in one HV subject, for different brain regions, including cerebellum, thalamus, striatum, and frontal and temporal cortex. Curves fitted with a 2-TC model are also shown

Table 1
Tracer binding outcome metrics corresponding to TACs in Fig. 5

Region	V_T (mL/mL)	BP_{ND}
Cerebellum	9.59	4.23
Thalamus	14.54	8.11
Striatum	11.21	5.26
Front ctx	9.16	5.08
Temp ctx	11.12	7.44

least-squares cost function. V_B can either be estimated as an additional parameter or can be fixed to some reasonable value (e.g., 4–5%).

Five TACs from a GE-179 study on an HV subject, corresponding to different brain regions, are shown in Fig. 5, together with curves fitted with a 2-TC model. In general, curves with higher initial amplitude correspond to regions with high blood flow, while curves with slower wash-out correspond to regions with higher tracer binding.

Due to relatively high levels of uncertainty in the estimation of the individual rate constants, various biologically meaningful outcome measures have been defined, which are derived from the rate constants, and can be determined with greater precision. Specific binding can be quantified by the ratio of $k_3{}'$ and k_4, known as the non-displaceable binding potential,

$$BP_{ND} = k_3'/k_4 \tag{5}$$

which is related to the true binding potential, defined as $BP = B_{avail}/K_D$ (or $BP = B_{max}/K_D$), where K_D is the dissociation constant, $K_D = k_{off}/k_{on}$ [59, 60].

Another useful outcome measure is the so called "total volume of distribution," defined as:

$$V_T = \frac{K_1}{k_2'}\left(1 + \frac{k_3'}{k_4}\right) \tag{6}$$

The name is an unfortunate choice and could be misleading, as the quantity in question does not actually represent a distribution volume but is a measure of tracer binding. This outcome measure incorporates both specific and nonspecific binding, as well as free tracer in tissue. V_T is useful as it reflects tracer binding while being independent of blood flow and can be determined with high precision.

If there is a reference region, i.e., a brain region devoid of specific binding, the binding potential can be derived as:

$$BP_{\text{ND}} = \frac{(V_T - V_R)}{V_R} = \frac{V_T}{V_R} - 1 \tag{7}$$

where V_R is the V_T values of the reference region. This assumes that the ND uptake is the same in both regions.

Estimated V_T and BP_{ND} values for the TACs shown in Fig. 5 are presented in Table 1. There is a larger relative variability in the BP_{ND} values, as these are more specific. Notice that, while the striatum and temporal cortex have similar V_T, the latter has significantly higher BP_{ND}. Also, the striatum has a higher V_T than the frontal cortex, but the two have similar BP_{ND}.

The 2-TC model is not always the most appropriate one to use. Sometimes a simple 1-TC model with two rate constants (K_1 and k_2'') may be good enough, and sometimes an irreversible model may be the best choice. In this case, you would set $k_4 = 0$, and the main outcome metric would be the influx rate:

$$K_i = K_1 \frac{k_3'}{k_2 + k_3'} \tag{8}$$

To find out which model is best, you can compare the fits of different models using, e.g., the Akaike information criteria (AIC), which takes into account the residuals of the fit, as well as the number of parameters in the model [61]. You should not add more parameters to a model if it does not significantly improve the fit (cf. Occam's razor).

3.3 Other Analysis Techniques

An alternative approach is to use the spectral analysis (SA) technique [62]. Here, a family of basis functions is first generated as the convolution of the AIF with a series of monoexponential functions. An algorithm will then automatically choose the smallest number of basis functions needed to describe the TAC, based on a nonnegative least-squares fit. The number of basis functions chosen can be interpreted as the number of compartments in an underlying model. (See (3) above for the 2-TC model.)

Graphical analysis is another simplified approach, which includes various techniques that are based on plotting transformed versions of the measured data. The Patlak plot is used for irreversible, and the Logan plot for reversible tracers [63, 64]. The latter technique, which is the most relevant one in the current context, is based on the equation:

$$\frac{\int_0^t C_T(\tau)d\tau}{C_T(t)} = V_T \frac{\int_0^t C_p(\tau)d\tau}{C_T(t)} + c \Longleftrightarrow Y(t) = V_T \cdot X(t) + c; t > t^* \tag{9}$$

where c is a constant and t^* is time required for the tracer to reach equilibrium.

By plotting $\Upsilon(t)$ vs. $X(t)$, a curve is obtained, which becomes a straight line with a slope equal to V_T. It is not necessary to actually plot the data to use this technique, as long as the t^* value is known.

3.4 Arterial Input Function

As mentioned above, an input function is needed in order to perform kinetic analysis; ideally, this would be the AIF, which typically has a sharp initial peak, followed by a slow tail. The gold standard procedure for obtaining the AIF involves arterial sampling. Sample intervals need to be short in the beginning (~5 to 10 s for ~90 s or so) but can then become longer and longer, with up to 5 min intervals toward the end.

The samples are centrifuged, and plasma is extracted. The activity concentration in whole blood (C_B) and plasma (C_a) are measured separately in a well counter. In a subset of samples (~6), some of the blood is used for metabolite analysis with high-pressure liquid chromatography (HPLC) (*see* **Note 6**). The parent fraction, f_{par}, vs. time is fitted with an analytical function [65]. The AIF is then obtained as:

$$C_p(t) = C_a(t) \cdot f_{par}(t) \tag{10}$$

The procedure of arterial sampling is invasive and not always practicable. Therefore, alternative methods have been developed. If there is a reference region, the TAC of this region can be used as an indirect input function. However, in the case of NMDAR, there is no reference region, as these receptors are present ubiquitously in the brain [32].

Another alternative is the image-derived input function (IDIF), obtained from a TAC corresponding to a blood pool within the tomographic images [66]. In brain studies, the carotid arteries have been used for this purpose [67, 68]. The main problem is the relatively small diameter (4–5 mM), which is comparable to the typical spatial resolution of PET scanners. This means that the data will be subject to PVE, and an appropriate correction (PVC) will be crucial. This can be done based on anatomical information from a co-registered MRI image. Venous samples can be taken for measurement of f_{par} and the plasma over whole-blood ratio, f_{pob}. The IDIF is then given by:

$$C_p(t) = C_B(t) \cdot f_{pob}(t) \cdot f_{par}(t) \tag{11}$$

An alternative method for estimating f_{par} in combination with an IDIF, without the need for blood samples, has been proposed, based on the simultaneous estimation method (SIME) [69, 70]. In this case, f_{pob} would be obtained from population data. The SIME by itself can in principle estimate the shape of the AIF from several different tissue TACs, but not the absolute scale. By combining SIME with an IDIF, a more accurate and quantitative AIF is obtained.

Finally, an alternative approach that allows for the estimation of V_T without arterial sampling is the use of a bolus/infusion protocol [38]. The basic idea is to reach an equilibrium situation, in which both tissue and blood TACs are constant. In that situation, V_T is given by the tissue-to-plasma concentration ratio ($V_T = [C_T/C_p]_{eq}$).

Arterial samples can be replaced by venous samples, and a correction for the arterial-to-venous concentration ratio can be applied based on population data. Also, if a perfect equilibrium is not reached, a slope correction may be required.

4 Notes

4.1 Radionuclides and Modalities (Note 1)

The PET tracers mentioned above are labeled with either ^{11}C ($T_{1/2}$ = 20 min) or ^{18}F ($T_{1/2} = 110$ min). The former has the advantage of easier chemistry but the disadvantages of a short half-life and the requirement of an on-site cyclotron. Two tracers (GMOM and CNS-5161) were initially labeled with ^{11}C for evaluation purposes, and later ^{18}F-labeled versions were developed, as this is more practical, allowing for longer scans as well as transportation of the tracer from a production site to a separate scanning site.

The SPECT tracers mentioned are all labeled with ^{123}I, which has an even longer half-life (13 h). This makes it easy to transport between sites and allows for long scanning times; however, the radiation dose per unit injected activity will be higher, which limits the tracer dose that can be administered.

PET is also advantageous as compared to SPECT due to its higher sensitivity and better spatial resolution, resulting in superior image quality and improved tracer uptake quantification.

4.2 Tracer Selection (Note 2)

Whether or not a radioligand is suitable for PET or SPECT imaging depends on a number of factors: (1) It should have a high affinity for the targeted binding site; (2) it should also have high selectivity for the target site compared to other binding sites; (3) it should not have high lipophilicity, as this leads to high nonspecific binding; and (4) fast metabolism is an undesirable property, as this result in fast clearance from blood and subsequently from the brain. During tracer development, these properties are determined in in vitro experiments and in preclinical studies prior to human studies.

4.3 Specific Binding Validation (Note 3)

It is of fundamental importance to demonstrate that a tracer really exhibits specific binding to the targeted binding site. The first step is to show blocking of the target by the unlabeled compound itself, and the next to show blocking by a compound that is known to bind selectively to the target site, e.g., MK-801 or ketamine for NMDAR$_{IC}$ [43]. This can be done by performing two scans, without and with pre-injection of the competing compound.

Alternatively, a displacement study could be done by the administration of the compound during the equilibrium phase of a continuous infusion study [39].

In the case of $NMDAR_{IC}$, blocking or displacement studies are not straightforward. Taking into consideration the physiological effects on the subject, these studies are preferably made on laboratory animals that are sedated or anesthetized to avoid motion. It has been argued that anesthesia can be a confounding factor, as some anesthetic agents have inhibitory effects on NMDARs [43]. It is believed that this was the reason why extensive preclinical experiments, including blocking studies with various compounds as well as activation with methamphetamine, failed to demonstrate modulation of the $[^{18}F]$GE-179 signal [71, 72].

An alternative to blocking or displacing a tracer is to perform an intervention to increase the binding, as was done by Vibholm et al., using deep brain stimulation in rats [52]. Yet another approach is to scan patients with a condition that is believed would lead to higher tracer uptake [51].

4.4 Specific Binding Estimation (Note 4)

In most cases, $NMDAR_{IC}$ tracer uptake is analyzed with a reversible model, and the primary outcome measure used is V_T. This quantity is robust, but not specific. Apart from free and nonspecifically bound tracers in tissue, V_T also depends on nonspecific binding in the blood (protein binding). This component is assumed to equilibrate quickly, and the free fraction in plasma (f_p) is treated as a constant. Although it can be measured with various techniques, this is not usually done (presumably due to low precision in these types of measurements). The common assumption is therefore that f_p is similar on average between different subject groups.

Had a reference region been available, it would have been possible to derive various forms of binding potential (BP_F, BP_P or BP_{ND}), more directly related to the specific binding of the tracer [60]. As an alternative, Pilowsky et al. introduced two "specific binding indices," defined as $BI_1 = V_T - V_R$ and $BI_2 = (V_T - V_R)/V_R$, where V_R was obtained from the whole cortex [42]. Although not optimal, these binding indices would control for f_{ND} and f_p, respectively, and this led to an important clinical finding.

4.5 Test– Retest (Note 5)

An important property of a tracer is the reproducibility. This is usually addressed with so-called test-retest studies, i.e., two separate scans on the same subject. It has been argued that, since the opening probability of NMDARs at rest is low and furthermore depends on the cognitive and emotional state of the subject, the results of test-retest studies could be unpredictable without appropriate control for these factors [50]. This may explain the results presented by van der Aart et al. [46].

4.6 Metabolites (Note 6)

During a dynamic scan, tracer metabolites become increasingly present in blood due to metabolism in peripheral organs, such as the liver or lungs. Metabolite correction of the AIF is performed as described above. However, it is assumed that the metabolites do not cross the BBB. In order to verify that this is the case, modeling experiments can be done with two input functions, with and without metabolites. The results would then be compared with those of standard modeling using AIC [44].

References

1. Watkins JC, Evans RH (1981) Excitatory amino acid transmitters. Annu Rev Pharmacol Toxicol 21:165–204. https://doi.org/10.1146/annurev.pa.21.040181.001121

2. Palmada M, Centelles JJ (1998) Excitatory amino acid neurotransmission. Pathways for metabolism, storage and reuptake of glutamate in brain. Front Biosci 3:d701–d718. https://doi.org/10.2741/a314

3. Mori H, Mishina M (1995) Structure and function of the NMDA receptor channel. Neuropharmacology 34(10):1219–1237. https://doi.org/10.1016/0028-3908(95)00109-j

4. Choi DW (1995) Calcium: still center-stage in hypoxic-ischemic neuronal death. Trends Neurosci 18(2):58–60

5. Weiss JH, Hartley DM, Koh J, Choi DW (1990) The calcium channel blocker nifedipine attenuates slow excitatory amino acid neurotoxicity. Science 247(4949 Pt 1):1474–1477. https://doi.org/10.1126/science.247.4949.1474

6. Hardingham GE, Bading H (2010) Synaptic versus extrasynaptic NMDA receptor signalling: implications for neurodegenerative disorders. Nat Rev Neurosci 11(10):682–696. https://doi.org/10.1038/nrn2911

7. Papouin T, Ladepeche L, Ruel J, Sacchi S, Labasque M, Hanini M, Groc L, Pollegioni L, Mothet JP, Oliet SH (2012) Synaptic and extrasynaptic NMDA receptors are gated by different endogenous coagonists. Cell 150(3):633–646. https://doi.org/10.1016/j.cell.2012.06.029

8. Lisman JE, Fellous JM, Wang XJ (1998) A role for NMDA-receptor channels in working memory. Nat Neurosci 1(4):273–275. https://doi.org/10.1038/1086

9. Wang XJ (1999) Synaptic basis of cortical persistent activity: the importance of NMDA receptors to working memory. J Neurosci 19(21):9587–9603. https://doi.org/10.1523/JNEUROSCI.19-21-09587.1999

10. Chang JP, Lane HY, Tsai GE (2014) Attention deficit hyperactivity disorder and N-methyl-D-aspartate (NMDA) dysregulation. Curr Pharm Des 20(32):5180–5185. https://doi.org/10.2174/13816128196661401110115227

11. Bubenikova-Valesova V, Horacek J, Vrajova M, Hoschl C (2008) Models of schizophrenia in humans and animals based on inhibition of NMDA receptors. Neurosci Biobehav Rev 32(5):1014–1023. https://doi.org/10.1016/j.neubiorev.2008.03.012

12. Magnusson KR, Brim BL, Das SR (2010) Selective vulnerabilities of N-methyl-D-aspartate (NMDA) receptors during brain aging. Front Aging Neurosci 2:11. https://doi.org/10.3389/fnagi.2010.00011

13. Bressan RA, Pilowsky LS (2000) Imaging the glutamatergic system in vivo–relevance to schizophrenia. Eur J Nucl Med 27(11):1723–1731. https://doi.org/10.1007/s002590000372

14. Al-Diwani A, Handel A, Townsend L, Pollak T, Leite MI, Harrison PJ, Lennox BR, Okai D, Manohar SG, Irani SR (2019) The psychopathology of NMDAR-antibody encephalitis in adults: a systematic review and phenotypic analysis of individual patient data. Lancet Psychiatry 6(3):235–246. https://doi.org/10.1016/S2215-0366(19)30001-X

15. Hughes EG, Peng X, Gleichman AJ, Lai M, Zhou L, Tsou R, Parsons TD, Lynch DR, Dalmau J, Balice-Gordon RJ (2010) Cellular and synaptic mechanisms of anti-NMDA receptor encephalitis. J Neurosci 30(17):5866–5875. https://doi.org/10.1523/JNEUROSCI.0167-10.2010

16. Moscato EH, Peng X, Jain A, Parsons TD, Dalmau J, Balice-Gordon RJ (2014) Acute mechanisms underlying antibody effects in anti-N-methyl-D-aspartate receptor encephalitis. Ann Neurol 76(1):108–119. https://doi.org/10.1002/ana.24195

17. Ferreira IL, Ferreiro E, Schmidt J, Cardoso JM, Pereira CM, Carvalho AL, Oliveira CR, Rego AC (2015) Abeta and NMDAR activation cause mitochondrial dysfunction involving ER calcium release. Neurobiol Aging 36(2): 680–692. https://doi.org/10.1016/j.neurobiolaging.2014.09.006

18. Talantova M, Sanz-Blasco S, Zhang X, Xia P, Akhtar MW, Okamoto S, Dziewczapolski G, Nakamura T, Cao G, Pratt AE, Kang YJ, Tu S, Molokanova E, McKercher SR, Hires SA, Sason H, Stouffer DG, Buczynski MW, Solomon JP, Michael S, Powers ET, Kelly JW, Roberts A, Tong G, Fang-Newmeyer T, Parker J, Holland EA, Zhang D, Nakanishi N, Chen HS, Wolosker H, Wang Y, Parsons LH, Ambasudhan R, Masliah E, Heinemann SF, Pina-Crespo JC, Lipton SA (2013) Abeta induces astrocytic glutamate release, extrasynaptic NMDA receptor activation, and synaptic loss. Proc Natl Acad Sci USA 110(27):E2518–E2527. https://doi.org/10.1073/pnas.1306832110

19. Ahmed I, Bose SK, Pavese N, Ramlackhansingh A, Turkheimer F, Hotton G, Hammers A, Brooks DJ (2011) Glutamate NMDA receptor dysregulation in Parkinson's disease with dyskinesias. Brain 134(Pt 4):979–986. https://doi.org/10.1093/brain/awr028

20. Fan MM, Raymond LA (2007) N-methyl-D-aspartate (NMDA) receptor function and excitotoxicity in Huntington's disease. Prog Neurobiol 81(5–6):272–293. https://doi.org/10.1016/j.pneurobio.2006.11.003

21. Borris DJ, Bertram EH, Kapur J (2000) Ketamine controls prolonged status epilepticus. Epilepsy Res 42(2–3):117–122. https://doi.org/10.1016/s0920-1211(00)00175-3

22. Modica PA, Tempelhoff R, White PF (1990) Pro- and anticonvulsant effects of anesthetics (part I). Anesth Analg 70(3):303–315. https://doi.org/10.1213/00000539-199003000-00013

23. Peltz G, Pacific DM, Noviasky JA, Shatla A, Mehalic T (2005) Seizures associated with memantine use. Am J Health Syst Pharm 62(4):420–421. https://doi.org/10.1093/ajhp/62.4.0420

24. Sveinbjornsdottir S, Sander JW, Upton D, Thompson PJ, Patsalos PN, Hirt D, Emre M, Lowe D, Duncan JS (1993) The excitatory amino acid antagonist D-CPP-ene (SDZ EAA-494) in patients with epilepsy. Epilepsy Res 16(2):165–174. https://doi.org/10.1016/0920-1211(93)90031-2

25. Ramaswamy P, Aditi Devi N, Hurmath Fathima K, Dalavaikodihalli Nanjaiah N

(2014) Activation of NMDA receptor of glutamate influences MMP-2 activity and proliferation of glioma cells. Neurol Sci 35(6): 823–829. https://doi.org/10.1007/s10072-013-1604-5

26. Waterhouse RN (2003) Imaging the PCP site of the NMDA ion channel. Nucl Med Biol 30(8):869–878. https://doi.org/10.1016/s0969-8051(03)00127-6

27. Galovic M, Koepp M (2016) Advances of molecular imaging in epilepsy. Curr Neurol Neurosci Rep 16(6):58. https://doi.org/10.1007/s11910-016-0660-7

28. Ransom RW, Eng WS, Burns HD, Gibson RE, Solomon HF (1990) (+)-3-[123I]Iodo-MK-801: synthesis and characterization of binding to the N-methyl-D-aspartate receptor complex. Life Sci 46(15):1103–1110. https://doi.org/10.1016/0024-3205(90)90420-v

29. Owens J, Wyper DJ, Patterson J, Brown DR, Elliott AT, Teasdale GM, McCulloch J (1997) First SPET images of glutamate (NMDA) receptor activation in vivo in cerebral ischaemia. Nucl Med Commun 18(2):149–158. https://doi.org/10.1097/00006231-199702000-00010

30. Brown DR, Wyper DJ, Owens J, Patterson J, Kelly RC, Hunter R, McCulloch J (1997) 123Iodo-MK-801: a spect agent for imaging the pattern and extent of glutamate (NMDA) receptor activation in Alzheimer's disease. J Psychiatr Res 31(6):605–619. https://doi.org/10.1016/s0022-3956(97)00031-9

31. Blin J, Denis A, Yamaguchi T, Crouzel C, MacKenzie ET, Baron JC (1991) PET studies of [18F]methyl-MK-801, a potential NMDA receptor complex radioligand. Neurosci Lett 121(1–2):183–186. https://doi.org/10.1016/0304-3940(91)90680-r

32. Hartvig P, Valtysson J, Antoni G, Westerberg G, Langstrom B, Ratti Moberg E, Oye I (1994) Brain kinetics of (R)- and (S)-[N-methyl-11C]ketamine in the rhesus monkey studied by positron emission tomography (PET). Nucl Med Biol 21(7):927–934. https://doi.org/10.1016/0969-8051(94)90081-7

33. Kumlien E, Hartvig P, Valind S, Oye I, Tedroff J, Langstrom B (1999) NMDA-receptor activity visualized with (S)-[N-methyl-11C]ketamine and positron emission tomography in patients with medial temporal lobe epilepsy. Epilepsia 40(1):30–37. https://doi.org/10.1111/j.1528-1157.1999.tb01985.x

34. Ametamey SM, Samnick S, Leenders KL, Vontobel P, Quack G, Parsons CG, Schubiger PA (1999) Fluorine-18 radiolabelling,

biodistribution studies and preliminary PET evaluation of a new memantine derivative for imaging the NMDA receptor. J Recept Signal Transduct Res 19(1–4):129–141. https://doi.org/10.3109/10799899909036640

35. Ametamey SM, Bruehlmeier M, Kneifel S, Kokic M, Honer M, Arigoni M, Buck A, Burger C, Samnick S, Quack G, Schubiger PA (2002) PET studies of 18F-memantine in healthy volunteers. Nucl Med Biol 29(2):227–231. https://doi.org/10.1016/s0969-8051(01)00293-1

36. Owens J, Tebbutt AA, McGregor AL, Kodama K, Magar SS, Perlman ME, Robins DJ, Durant GJ, McCulloch J (2000) Synthesis and binding characteristics of N-(1-naphthyl)-N'-(3-[(125)I]-iodophenyl)-N'-methylguani-dine ([(125)I]-CNS 1261): a potential SPECT agent for imaging NMDA receptor activation. Nucl Med Biol 27(6):557–564. https://doi.org/10.1016/s0969-8051(00)00102-5

37. Erlandsson K, Bressan RA, Mulligan RS, Gunn RN, Cunningham VJ, Owens J, Wyper D, Ell PJ, Pilowsky LS (2003) Kinetic modelling of [123I]CNS 1261 – a potential SPET tracer for the NMDA receptor. Nucl Med Biol 30(4):441–454. https://doi.org/10.1016/s0969-8051(02)00450-x

38. Bressan RA, Erlandsson K, Mulligan RS, Gunn RN, Cunningham VJ, Owens J, Cullum ID, Ell PJ, Pilowsky LS (2004) A bolus/infusion paradigm for the novel NMDA receptor SPET tracer [123I]CNS 1261. Nucl Med Biol 31(2):155–164. https://doi.org/10.1016/j.nucmedbio.2003.08.008

39. Stone JM, Erlandsson K, Arstad E, Bressan RA, Squassante L, Teneggi V, Ell PJ, Pilowsky LS (2006) Ketamine displaces the novel NMDA receptor SPET probe [(123)I]CNS-1261 in humans in vivo. Nucl Med Biol 33(2):239–243. https://doi.org/10.1016/j.nucmedbio.2005.12.001

40. Stone JM, Erlandsson K, Arstad E, Squassante L, Teneggi V, Bressan RA, Krystal JH, Ell PJ, Pilowsky LS (2008) Relationship between ketamine-induced psychotic symptoms and NMDA receptor occupancy: a [(123)I]CNS-1261 SPET study. Psychopharmacology 197(3):401–408. https://doi.org/10.1007/s00213-007-1047-x

41. Bressan RA, Erlandsson K, Stone JM, Mulligan RS, Krystal JH, Ell PJ, Pilowsky LS (2005) Impact of schizophrenia and chronic antipsychotic treatment on [123I]CNS-1261 binding to N-methyl-D-aspartate receptors in vivo. Biol Psychiatry 58(1):41–46. https://doi.org/10.1016/j.biopsych.2005.03.016

42. Pilowsky LS, Bressan RA, Stone JM, Erlandsson K, Mulligan RS, Krystal JH, Ell PJ (2006) First in vivo evidence of an NMDA receptor deficit in medication-free schizophrenic patients. Mol Psychiatry 11(2):118–119. https://doi.org/10.1038/sj.mp.4001751

43. Waterhouse RN, Slifstein M, Dumont F, Zhao J, Chang RC, Sudo Y, Sultana A, Balter A, Laruelle M (2004) In vivo evaluation of [11C]N-(2-chloro-5-thiomethylphenyl)-N'-(3-methoxy-phenyl)-N'-methylguanidine ([11C]GMOM) as a potential PET radiotracer for the PCP/NMDA receptor. Nucl Med Biol 31(7):939–948. https://doi.org/10.1016/j.nucmedbio.2004.03.012

44. van der Doef TF, Golla SSV, Klein PJ, Oropeza-Seguias GM, Schuit RC, Metaxas A, Jobse E, Schwarte LA, Windhorst AD, Lammertsma AA, van Berckel BNM, Boellaard R (2016) Quantification of the novel N-methyl-d-aspartate receptor ligand [11C]GMOM in man. J Cereb Blood Flow Metab 36(6):1111–1121. https://doi.org/10.1177/0271678X15608391

45. Golla SS, Klein PJ, Bakker J, Schuit RC, Christiaans JA, van Geest L, Kooijman EJ, Oropeza-Seguias GM, Langermans JA, Leysen JE, Boellaard R, Windhorst AD, van Berckel BN, Metaxas A (2015) Preclinical evaluation of [(18)F]PK-209, a new PET ligand for imaging the ion-channel site of NMDA receptors. Nucl Med Biol 42(2):205–212. https://doi.org/10.1016/j.nucmedbio.2014.09.006

46. van der Aart J, Golla SSV, van der Pluijm M, Schwarte LA, Schuit RC, Klein PJ, Metaxas A, Windhorst AD, Boellaard R, Lammertsma AA, van Berckel BNM (2018) First in human evaluation of [(18)F]PK-209, a PET ligand for the ion channel binding site of NMDA receptors. EJNMMI Res 8(1):69. https://doi.org/10.1186/s13550-018-0424-2

47. Luthra S, Hume SP, Turton DR, Yongjun Z, Brady F, Hirani E, Ahmad R, Osman S, Fisher A, Smith T, Brooks DJ (2004) Radiosynthesis of [11C]CNS 5161 and preliminary evaluation as a potential PET ligand for the NMDA receptor. NeuroImage 22:T154–T155

48. Asselin M-C, Hammers A, Turton DR, Osman S, Koepp MJ, Brooks DJ (2004) Initial kinetic analyses of the in vivo binding of the putative NMDA receptor ligand [C-11]CNS 5161 in humans. NeuroImage 22:T137–T138

49. Hammers A, Asselin M-C, Brooks DJ, Luthra SK, Hume SP, Thompson PJ, Turton DR, Duncan JS, Koepp MJ (2004) Correlation of memory function with binding of [C-11]CNS

5161, a novel putative NMDA ion channel PET ligand. NeuroImage 22:T54–T55

50. McGinnity CJ, Hammers A, Riano Barros DA, Luthra SK, Jones PA, Trigg W, Micallef C, Symms MR, Brooks DJ, Koepp MJ, Duncan JS (2014) Initial evaluation of 18F-GE-179, a putative PET tracer for activated N-methyl D-aspartate receptors. J Nucl Med 55(3): 423–430. https://doi.org/10.2967/jnumed. 113.130641

51. McGinnity CJ, Koepp MJ, Hammers A, Riano Barros DA, Pressler RM, Luthra S, Jones PA, Trigg W, Micallef C, Symms MR, Brooks DJ, Duncan JS (2015) NMDA receptor binding in focal epilepsies. J Neurol Neurosurg Psychiatry 86(10):1150–1157. https://doi.org/10. 1136/jnnp-2014-309897

52. Vibholm AK, Landau AM, Moller A, Jacobsen J, Vang K, Munk OL, Orlowski D, Sorensen JC, Brooks DJ (2021) NMDA receptor ion channel activation detected in vivo with [(18)F]GE-179 PET after electrical stimulation of rat hippocampus. J Cereb Blood Flow Metab 41(6):1301–1312. https://doi.org/10. 1177/0271678X20954928

53. Beck K, Arumuham A, Veronese M, Santangelo B, McGinnity CJ, Dunn J, McCutcheon RA, Kaar SJ, Singh N, Pillinger T, Borgan F, Stone J, Jauhar S, Sementa T, Turkheimer F, Hammers A, Howes OD (2021) N-methyl-D-aspartate receptor availability in first-episode psychosis: a PET-MR brain imaging study. Transl Psychiatry 11(1):425. https://doi.org/10.1038/ s41398-021-01540-2

54. Galovic M, Al-Diwani A, Vivekananda U, Walker MC, Irani SR, Koepp MJ, Investigators N (2023) In vivo N-methyl-d-aspartate receptor (NMDAR) density as assessed using positron emission tomography during recovery from NMDAR-antibody encephalitis. JAMA Neurol 80(2):211–213. https://doi.org/10. 1001/jamaneurol.2022.4352

55. Khalil MME (2021) Basic sciences of nuclear medicine. Springer

56. Gunn RN, Gunn SR, Cunningham VJ (2001) Positron emission tomography compartmental models. J Cereb Blood Flow Metab 21(6): 635–652. https://doi.org/10.1097/ 00004647-200106000-00002

57. Koeppe RA, Holthoff VA, Frey KA, Kilbourn MR, Kuhl DE (1991) Compartmental analysis of [11C]flumazenil kinetics for the estimation of ligand transport rate and receptor distribution using positron emission tomography. J

Cereb Blood Flow Metab 11(5):735–744. https://doi.org/10.1038/jcbfm.1991.130

58. Kerwin RW, Pilowsky LS (1995) Traditional receptor theory and its application to neuroreceptor measurements in functional imaging. Eur J Nucl Med 22(7):699–710. https://doi. org/10.1007/BF01254574

59. Mintun MA, Raichle ME, Kilbourn MR, Wooten GF, Welch MJ (1984) A quantitative model for the in vivo assessment of drug binding sites with positron emission tomography. Ann Neurol 15(3):217–227. https://doi.org/10. 1002/ana.410150302

60. Innis RB, Cunningham VJ, Delforge J, Fujita M, Gjedde A, Gunn RN, Holden J, Houle S, Huang SC, Ichise M, Iida H, Ito H, Kimura Y, Koeppe RA, Knudsen GM, Knuuti J, Lammertsma AA, Laruelle M, Logan J, Maguire RP, Mintun MA, Morris ED, Parsey R, Price JC, Slifstein M, Sossi V, Suhara T, Votaw JR, Wong DF, Carson RE (2007) Consensus nomenclature for in vivo imaging of reversibly binding radioligands. J Cereb Blood Flow Metab 27(9):1533–1539. https://doi.org/10.1038/sj.jcbfm.9600493

61. Akaike H (1974) A new look at the statistical model identification. IEEE Trans Autom Control 19(6):716–723

62. Cunningham VJ, Jones T (1993) Spectral analysis of dynamic PET studies. J Cereb Blood Flow Metab 13(1):15–23. https://doi.org/ 10.1038/jcbfm.1993.5

63. Patlak CS, Blasberg RG, Fenstermacher JD (1983) Graphical evaluation of blood-to-brain transfer constants from multiple-time uptake data. J Cereb Blood Flow Metab 3(1):1–7. https://doi.org/10.1038/jcbfm.1983.1

64. Logan J, Fowler JS, Volkow ND, Wolf AP, Dewey SL, Schlyer DJ, MacGregor RR, Hitzemann R, Bendriem B, Gatley SJ et al (1990) Graphical analysis of reversible radioligand binding from time-activity measurements applied to [N-11C-methyl]-(-)-cocaine PET studies in human subjects. J Cereb Blood Flow Metab 10(5):740–747. https://doi. org/10.1038/jcbfm.1990.127

65. Tonietto M, Rizzo G, Veronese M, Fujita M, Zoghbi SS, Zanotti-Fregonara P, Bertoldo A (2016) Plasma radiometabolite correction in dynamic PET studies: insights on the available modeling approaches. J Cereb Blood Flow Metab 36(2):326–339. https://doi.org/10. 1177/0271678X15610585

66. Zanotti-Fregonara P, Fadaili el M, Maroy R, Comtat C, Souloumiac A, Jan S, Ribeiro MJ, Gaura V, Bar-Hen A, Trebossen R (2009)

Comparison of eight methods for the estimation of the image-derived input function in dynamic [(18)F]-FDG PET human brain studies. J Cereb Blood Flow Metab 29(11): 1825–1835. https://doi.org/10.1038/jcbfm. 2009.93

67. Sari H, Erlandsson K, Law I, Larsson HB, Ourselin S, Arridge S, Atkinson D, Hutton BF (2017) Estimation of an image derived input function with MR-defined carotid arteries in FDG-PET human studies using a novel partial volume correction method. J Cereb Blood Flow Metab 37(4):1398–1409. https://doi.org/10.1177/0271678X16656197

68. Galovic M, Erlandsson K, Fryer TD, Hong YT, Manavaki R, Sari H, Chetcuti S, Thomas BA, Fisher M, Sephton S, Canales R, Russell JJ, Sander K, Arstad E, Aigbirhio FI, Groves AM, Duncan JS, Thielemans K, Hutton BF, Coles JP, Koepp MJ, Investigators N (2021) Validation of a combined image derived input function and venous sampling approach for the quantification of [(18)F]GE-179 PET binding in the brain. NeuroImage 237:118194. https://doi.org/10.1016/j.neuroimage. 2021.118194

69. Sari H, Erlandsson K, Marner L, Law I, Larsson HBW, Thielemans K, Ourselin S, Arridge S, Atkinson D, Hutton BF (2018) Non-invasive kinetic modelling of PET tracers

with radiometabolites using a constrained simultaneous estimation method: evaluation with (11)C-SB201745. EJNMMI Res 8(1): 58. https://doi.org/10.1186/s13550-018-0412-6

70. Feng D, Wong KP, Wu CM, Siu WC (1997) A technique for extracting physiological parameters and the required input function simultaneously from PET image measurements: theory and simulation study. IEEE Trans Inf Technol Biomed 1(4):243–254. https://doi.org/10.1109/4233.681168

71. Schoenberger M, Schroeder FA, Placzek MS, Carter RL, Rosen BR, Hooker JM, Sander CY (2018) In vivo [(18)F]GE-179 brain signal does not show NMDA-specific modulation with drug challenges in rodents and nonhuman primates. ACS Chem Neurosci 9(2):298–305. https://doi.org/10.1021/acschemneuro. 7b00327

72. McGinnity CJ, Arstad E, Beck K, Brooks DJ, Coles JP, Duncan JS, Galovic M, Hinz R, Hirani E, Howes OD, Jones PA, Koepp MJ, Luo F, Riano Barros DA, Singh N, Trigg W, Hammers A (2019) Comment on "in vivo [(18)F]GE-179 brain signal does not show NMDA-specific modulation with drug challenges in rodents and nonhuman primates". ACS Chem Neurosci 10(1):768–772. https://doi.org/10.1021/acschemneuro. 8b00246

Chapter 2

Manipulation and Imaging of Plasticity of the Glutamate Synapse for Brain Diseases in Living Human

Takuya Takahashi

Abstract

Numerous numbers of synapses exist in the human brain to convey information from one neuron to another. Among them, excitatory glutamate synapses are major synapses in the brain. Glutamate α-amino-3-hydroxy-5-methyl-4-isoxazole propionic acid (AMPA) receptors play pivotal roles at glutamatergic synapses. Upon the stimulation of presynaptic neurons, glutamate is released from the presynaptic terminals and binds to glutamate receptors at the postsynaptic site. The AMPA receptor complex forms an ion channel, and the binding of glutamate activates AMPA receptors, and positively charged cations flow into the cells through AMPA receptors, which depolarizes the cell membrane. Physiological roles of AMPA receptors have been well characterized in experimental animals. Especially, synaptic trafficking of AMPA receptors has been recognized as a fundamental molecular mechanism of synaptic plasticity. However, these molecular dynamics of AMPA receptors in the living human brain remain to be elucidated. In this chapter, the development and the application of our recently developed synaptic plasticity enhancer small compound and novel PET (positron emission tomography) tracer, $[^{11}C]K$-2, the first technology to visualize and quantify AMPA receptors in the living human brain, will be discussed.

Key words Synaptic plasticity, AMPA receptor, Trafficking, PET, [11C]K-2, Edonerpic maleate, Stroke, Rehabilitation, Epilepsy

1 Introduction

More than 100 billion neurons form the adult human brain. The synapse is the structure to transmit information from a neuron to another. Upon the propagation of action potential to the presynaptic terminal, a chemical neurotransmitter (glutamate, GABA (gamma-aminobutyric acid), dopamine, etc.) stored in the presynaptic vesicles is released by exocytosis into the synaptic cleft and binds to receptors at the postsynaptic membrane, which leads to the transmission of neuronal information. Thousands of synapses interconnect neurons in the brain. Excitatory glutamate synapses in which the binding of glutamate to postsynaptic receptors induces elevation of postsynaptic membrane potential (depolarization). Inhibitory gamma-aminobutyric acid (GABA) synapses where

Daichi Sone (ed.), *Molecular Imaging for Brain Diseases*, Neuromethods, vol. 222, https://doi.org/10.1007/978-1-0716-4494-2_2,
© The Author(s), under exclusive license to Springer Science+Business Media, LLC, part of Springer Nature 2025

GABA binding to its receptors induces the decrease of postsynaptic membrane potential (hyperpolarization). These types of synapses mainly mediate rapid neurotransmission. [1]

Neuropsychiatric disorders severely impair the quality of life of patients. Biological mechanisms underlying these disorders, such as their onset, progression, and recovery process, are still poorly understood, which hampers the development of novel diagnostics and therapeutics. Recent studies from animal disease models, genetics, and postmortem brains suggest that malfunctions of synapses underlie these disorders.

Glutamate synapses are the major excitatory synapses in the brain and mediate a large part of neuronal functions. α-amino-3-hydroxy-5-methyl-4-isoxazole propionic acid receptor (AMPAR) is an ionotropic glutamate receptor. Binding of glutamate activates AMPARs, that induces the influx of cations, elevating postsynaptic membrane potential. There are four subunits of AMPARs, GluA1, GluA2, GluA3, and GluA4, which form a core functional ion channel [2, 3]. GluA1-3 are expressed in the large fraction of neurons in the nervous system of matured animals [4]. GluA4 is primarily expressed early in development [4]. N-methyl-D-aspartate receptor (NMDAR) is another ionotropic glutamate receptor. At the resting potential, NMDAR is blocked by the magnesium, which is released by the depolarization by the AMPAR activation leading to the influx of cations including calcium. Since AMPAR-mediated current is a major fraction of the glutamatergic synaptic current [5], AMPAR can be a promising translational target to overcome neuropsychiatric disorders.

2   AMPAR Trafficking and Experience-Dependent Synaptic Plasticity

Experience, such as learning and sensory inputs, changes neural circuits from development to adulthood, leading to the large diversity of neural functions in the brain, which is known as "neural plasticity." These circuit alterations depend on the input-specific strengthening and weakening of synaptic efficacy in the brain. The most studied forms of synaptic plasticity are long-term potentiation (LTP) and long-term depression (LTD), in which a short period of repetitive synaptic stimulation can induce long-lasting strengthening (LTP) or weakening (LTD) of synaptic transmission [6, 7]. Postsynaptic insertion or removal of AMPAR underlies LTP and LTD [2, 3, 8]. Synaptic trafficking and the addition of AMPAR is one of the most characterized synaptic molecular modifications underlying experience-dependent synaptic plasticity [9–13]. This molecular modification was first reported as a molecular mechanism of long-lasting strengthening of synaptic responses during LTP using in vitro hippocampal slice cultures [14]. In the hippocampal pyramidal synapses of matured rats, GluA1/2 and

GluA2/3 heteromers are the dominant configuration of AMPARs [15]. These subunit compositions are considered to be largely maintained in the brain, while the relative expression of AMPAR subunits exhibits some regional diversity [16, 17]. Subunit composition is thought to rule synaptic AMPAR trafficking [15]. GluA1-containing AMPARs are delivered and additionally inserted into synapses during LTP in an activity-dependent manner, which leads to the increase of AMPAR-mediated synaptic current [14, 18]. GluA1 homomeric receptors, Ca^{2+} permeable AMPAR compositions, are primarily delivered into synapses in the early phase of LTP [19]. Synaptic delivery of GluA2/3 heteromers is constitutive and activity-independent, which constitutes the basal synaptic transmission, and GluA2/3 heteromers are thought to replace GluA1-containing AMPARs and maintain the increased synaptic transmission during LTP [15]. The synaptic AMPAR delivery in experience-dependent neural plasticity has first been reported in the developing rat barrel cortex, which is responsible for the processing of sensory input from whiskers [9]. By combining viral-mediated in vivo gene transfer and electrophysiological whole-cell recording, we demonstrated that GluA1-containing AMPARs are trafficked into synapses formed from layer 4 to layer 2/3 pyramidal neurons of the developing rat barrel cortex (2 weeks after birth) [9]. The delivery was blocked without whisker inputs. As was demonstrated in LTP, GluA1-lacking (GluA2/3) AMPARs were trafficked into synapses in an experience-independent fashion. The study exhibited that, in the developing rat barrel cortex, whisker inputs drive GluA1-containing AMPARs into synapses which leads to the potentiation of synaptic efficacy [9]. This molecular event is followed by the replacement with GluA2/3, whose experience-independent constitutive synaptic delivery maintains elevated synaptic efficacy. Synaptic delivery of AMPARs has been demonstrated to underlie experience-dependent plasticity in other brain regions such as the amygdala and hippocampus [10, 12]. Auditory fear learning drives GluA1-containing AMPARs into amygdala synapses [10]. Further, prevention of synaptic trafficking of GluA1-containing AMPARs in amygdala by the overexpression of cytoplasmic portion of GluA1 (GluA1-ctail), which blocks synaptic trafficking of GluA1-containing AMPARs by potentially trapping proteins required for trafficking, aggravated the formation of auditory fear learning [10]. This suggests that GluA1-containing AMPAR trafficking in the amygdala is necessary for fear learning. LTP-like synaptic potentiation in the hippocampus was reported in the inhibitory avoidance task [20]: IA task, a hippocampus-dependent contextual fear learning. In this task, rodents can cross from a lightbox to a dark box, where an electric foot shock is given. As a result, animals learn to avoid the dark box and stay in a light box, which they do not naturally prefer. The latency to re-enter the dark box is prolonged when the animal successfully establishes IA fear learning. We found that IA learning induces synaptic delivery of GluA1-containing AMPARs into hippocampal CA3-CA1 synapses [12]. In addition, the expression of GluA1-ctail in the dorsal hippocampus prevented IA learning formation, which indicates that synaptic trafficking of GluA1-containing AMPARs is required for hippocampus-dependent learning [12]. Do synaptically presented AMPARs encode memory? To prove this hypothesis, we generated the monoclonal antibody to GluA1 (named Z9139) and labeled the antibody with light-sensitive eosin [21]. This procedure enables us to inactivate GluA1 by the light exposure in vitro and in vivo through chromophore-assisted light inactivation (CALI), which can perturb the functions of the target molecule by a photosensitizer producing short-lived reactive oxygen upon irradiation with the light [22]. CALI with eosin-labeled Z9139 specifically inactivated GluA1 homomeric AMPARs. GluA1 homomeric AMPARs display conductance at negative but not at positive potentials (so-called inward rectification), while GluA2-containing AMPARs such as GluA1/2 and GluA2/3 heteromers possess conductance at both negative and positive potentials [9, 15, 19]. We detected increased inward rectification at hippocampal CA3-CA1 synapses within 2 h after IA learning, which was not observed 24 h after conditioning [21]. This indicates that IA learning transiently drives GluA1 homomeric AMPARs into CA3-CA1 synapses. We injected eosin-labeled Z9139 into the dorsal hippocampus with the implantation of a light cannula and conditioned these animals with IA learning. The light exposure to the dorsal hippocampus of IA-conditioned animals, which were injected with eosin-labeled Z9139, decreased the latency to re-enter the dark box [21]. This suggested that disruption of GluA1 homomeric AMPARs erases fear memory. Therefore, synaptically presented GluA1 homomeric AMPARs with IA encode fear memory. These results demonstrated that synaptic AMPAR trafficking is a fundamental molecular mechanism underlying experience-dependent neuronal plasticity in the brain. Thus, the trafficking of AMPARs and, moreover, AMPARs themselves can be a promising translational target.

3   Synaptic AMPAR Trafficking as a Translational Target to Augment Functional Recovery

Functional recovery from brain damage such as stroke is thought to be a plastic neuronal process with functional compensation in intact brain areas [23]. We have previously reported that visual deprivation induces synaptic trafficking of AMPARs into layer 4 to layer 2/3 pyramidal synapses in the juvenile barrel cortex, which led to the functional improvement of the whisker-barrel systems, which could help animals with visual loss to capture spatial information with their whiskers (cross-modal compensation) [11]. Further, rehabilitation in hemiplegia promotes functional reorganization in intact peri-lesional regions. Therefore, we consider that synaptic AMPAR trafficking can be a translational target to enhance functional recovery after brain damage such as stroke. For the recovery of motor function in hemiplegic patients, a specific activation of motor circuits is required. In this section, we introduce our recent identification of a small compound, edonerpic maleate (also called T817), which facilitates motor function recovery after traumatic damage of the central nervous system such as stroke by the enhancement of synaptic AMPAR trafficking [24]. After in vitro screening of the library of compounds to modify neuronal properties, we focused on edonerpic maleate (T-817MA) to examine its effect on experience-dependent synaptic AMPAR trafficking [24]. As was described above, whisker inputs drive GluA1-containing AMPARs into synapses formed from layer 4 to layer 2/3 pyramidal neurons of developing rodent barrel cortex [9]. This natural experience-driven synaptic GluA1-containing AMPAR trafficking terminates by juvenile age when the barrel cortex is functionally matured [11]. We administered edonerpic maleate to adult mice and investigated synaptic efficacy at layer 4-2/3 pyramidal synapses with whole-cell recordings of acute brain slices. We detected an increased AMPAR-mediated synaptic current at these synapses of animals administered with edonerpic maleate compared to vehicle-treated animals [24]. The increase of AMPAR-mediated synaptic current in animals treated with edonerpic maleate was abolished without whiskers, suggesting that edonerpic maleate facilitates synaptic AMPAR (GluA1-containing AMPAR, see also below) trafficking in an experience-dependent fashion [24]. A target protein of edonerpic maleate appeared to be CRMP2 (collapsin-response-mediator-protein 2), which was initially identified as a downstream molecule of semaphorin [25]. The semaphorin family was first isolated as a chemorepellent of developing axons [26, 27] and later reported to regulate synaptic functions [28]. In accordance with these findings, CRMP2 was also reported to regulate dendritic spine morphology [29–31]. The facilitation of experience-dependent synaptic AMPAR delivery with edonerpic maleate was not detected in CRMP2 knock-out mice, showing that the effect of edonerpic maleate on synaptic AMPAR delivery is mediated by the binding to CRMP2 [24]. Dephosphorylated (activated) actin depolarizing factor (ADF)/cofilin increases actin turnover and enhances AMPAR delivery during LTP [32]. Edonerpic maleate–induced facilitation of AMPAR trafficking was prevented by the expression of the dominant-negative form of ADF/cofilin [24]. Thus, the edonerpic maleate/CRMP2 complex increases dephosphorylated (activated) ADF/cofilin, which leads to enhancement of AMPAR delivery [24]. Based on the pharmacological action of edonerpic maleate in the brain, we applied edonerpic maleate to acute brain injury models to examine if it accelerates rehabilitative-training-dependent functional recovery. In the rodent model, mice were administered edonerpic maleate or vehicle for 21 days after cryogenic injury of the motor cortex. In this experiment, the group of mice with edonerpic maleate administration in combination with rehabilitative training showed greater improvements in the reach-to-grasp tasks compared to the control group of mice. Interestingly, mice with edonerpic maleate administration without rehabilitative training exhibited no improvements [24]. These results suggested that edonerpic maleate accelerates motor function recovery in a training-dependent manner. Mice administered edonerpic maleate 3 days after cryogenic injury also exhibited motor function recovery [24]. Edonerpic maleate–induced augmentation of functional recovery was accompanied by the increase of AMPAR-mediated synaptic currents in the peri-injured brain region and was blocked by the overexpression of GluA1-ctail in these areas, suggesting that edonerpic maleate augments training-dependent synaptic delivery of GluA1-containing AMPARs, leading to the enhancement of functional recovery with edonerpic maleate [24]. Consistent with the electrophysiological experiments, the effect of edonerpic maleate on the functional recovery was prevented in CRMP2 knockout mice, indicating that this effect is mediated by the binding of edonerpic maleate to CRMP2 [24]. In nonhuman primate model with internal capsule hemorrhage, *Macaca fascicularis* administered with edonerpic maleate after the induction of internal capsule hemorrhage also exhibited a more drastic recovery in dexterity tasks than the control animals [24]. These trans-species effects of edonerpic maleate for functional recovery should be emphasized to translate the result of the basic study to clinical trials.

4   Development of Positron Emission Tomography (PET) Tracer to Visualize and Quantify the Density of AMPAR in Living Human Brain

As mentioned above, it is important to investigate the density of AMPAR in the living human brain to elucidate the biological basis of neuropsychiatric disorders with the goal of developing novel diagnostics and therapeutics based on tangible biological evidence. Thus, a technology to visualize and quantify AMPARs in the living human brain has long been desired. However, considerable previous efforts have been unsuccessful [1]. PET tracers for AMPAR have been designed and radiolabeled upon AMPAR-binding characteristics such as antagonist and potentiator [33]. These are also subdivided into competitive or noncompetitive to glutamate. Compared with noncompetitive antagonists, fewer tracers based on potentiators were previously reported [34]. For example, [^{11}C] HMS011, synthesized from a noncompetitive antagonist of AMPAR, reached clinical evaluation [35, 36]. However, this tracer was reported to not be suitable for practical PET imaging in humans due to a small dynamic range in the brain, that could be due to its pharmacodynamic behavior as a substrate of P-gp, a main efflux transporter expressed in blood–brain barrier (BBB) [36]. In addition, other AMPAR antagonists have been developed as PET tracers; however, the development of these tracers was stopped during preclinical evaluation for the following reasons: (1) the binding patterns in the brain were likely the mixture of nonspecific binding and regional blood flow, (2) brain uptakes were too low to evaluate the binding potential in animals, and (3) no specific bindings were detected with PET imaging in animals due to the low affinity in vivo [37, 38]. AMPAR potentiators also have the same problems in their development, which exhibited a rapid decrease in the brain content, a uniform distribution, and high nonspecific binding due to their low affinity. In summary, an available PET tracer to quantify AMPAR in the living human brain requires outstanding properties such as high BBB penetration, low nonspecific binding, and high binding potential.

We have recently developed the first PET probe for AMPARs, [^{11}C]K-2, which enables us to visualize and quantify the AMPAR density in the living human brain [39]. As candidate compounds of PET tracers for AMPARs, we focused on noncompetitive positive allosteric modulators (PAMs), which do not compete with glutamate for the binding to AMPARs. Based on the chemical structure and affinity, we selected 4-[2-(phenylsulfonyl amino)ethylthio]-2,6-difluoro-phenoxyacetamide (PEPA) for radiolabeling with a [^{11}C]methyl group. We designed an ^{11}C-labeled derivative of PEPA and finally produced [^{11}C]K-2 [39]. Using in vitro and in vivo preclinical experiments, we accumulated evidence that [^{11}C]K-2 exhibited specific binding to AMPARs as described below: (1) K-2 application potentiated AMPAR-mediated synaptic current using acute hippocampal slices at a therapeutic dose as was observed with PEPA, (2) [^{11}C]K-2 uptake was decreased in the striatum where AMPAR expression was knocked down with a short hairpin RNA targeted to AMPARs, and (3) off-target binding assay proved that K-2 does not bind to major 160 proteins expressed in the central nervous system [39]. Furthermore, PET imaging of healthy subjects with [^{11}C]K-2 confirmed its reversible binding using Logan graphical analysis [39]. We also performed a [^{11}C]K-2 PET imaging to measure AMPAR density in patients with mesial temporal lobe epilepsy who planned to undergo surgery to resect the epileptogenic focus and then quantified AMPAR protein distribution in surgical specimens. The uptake was in a significant correlation with the local AMPAR protein distribution in the resected surgical specimens from the same individuals [39]. These results demonstrated that [^{11}C]K-2 is a potent PET tracer for AMPARs. Furthermore, PET images obtained from [^{11}C]K-2 turned out to represent AMPAR on the cell surface, a physiologically crucial fraction of AMPAR [40]. In addition, the safety of [^{11}C]K-2 for the application to human was confirmed by the phase I clinical trial [41].

5   Elucidation of the Biology of Epileptic Brain with [^{11}C]K-2

Presurgical identification of the epileptogenic zone is crucial to control seizures after surgical resection in patients with epilepsy. Using [^{11}C]K-2, we PET-imaged patients with epilepsy who underwent surgery after examination with intracranial electroencephalogram (MEG). The location of the dipole sources that best fit the measured magnetic fields was calculated using a single equivalent current dipole (ECD) model at the peak of each spike. We detected high [^{11}C]K-2 uptake in brain areas with ECDs compared to the other brain hemisphere, where we did not observe current dipoles (Fig. 1) [42]. This suggests that [^{11}C]K-2 imaging can delineate the epileptogenic zone.

Next, 30 patients with epilepsy and 31 healthy controls were scanned using PET with [^{11}C]K-2 which measures the density of cell surface AMPAR [43]. In patients with focal onset seizures, an

AMPA receptor levels are increased in the regions where ECDs are presumed by MEG

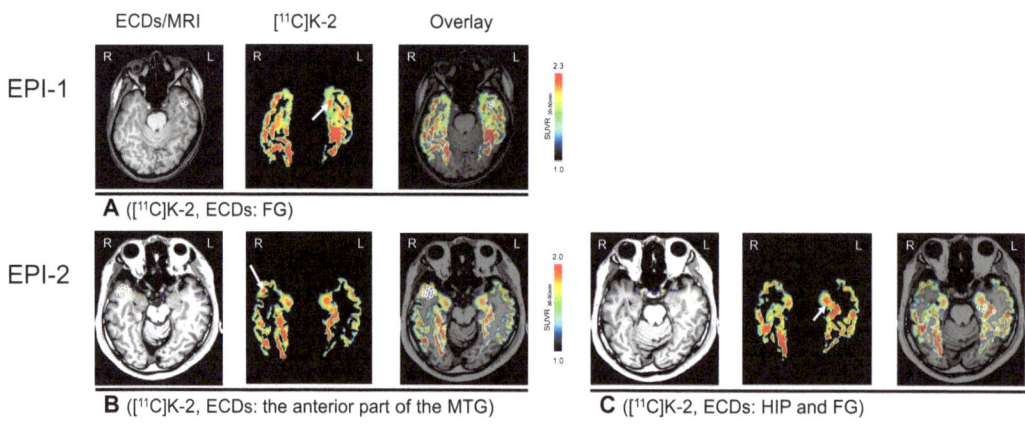

Fig. 1 Increased [^{11}C]K-2 uptake colocalizes with ECDs in the temporal lobes in patients with mesial temporal lobe epilepsy. Representative [^{11}C]K-2 images are shown restricted to temporal lobes where ECDs are estimated. Each panel shows ECDs on MRI (left), [^{11}C]K-2 (middle), and overlay (right). White arrows indicate the area showing elevated [^{11}C]K-2 uptakes than the other hemisphere. White circles indicate estimated ECDs. FG: fusiform gyrus, MTG: the middle temporal gyrus, HIP: the hippocampus. (Derived from Miyazaki et al. [42])

Hebbian and homeostatic plasticity functions in pathological conditions of epileptic brain

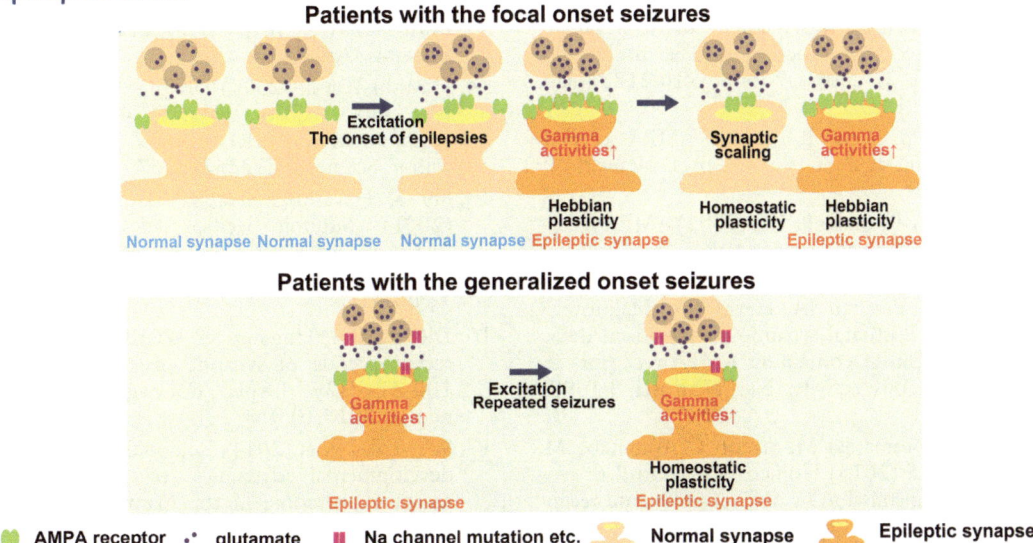

Fig. 2 Potential synaptic mechanisms underlying epilepsy. In partial epilepsy, strengthened synapses potentially due to Hebbian plasticity can build up epileptic circuits, while physiologically normal synapses can be downregulated by synaptic scaling (homeostatic plasticity). In the generalized onset seizure, down-regulation of synaptic AMPAR due to synaptic scaling can contribute to suppressing the overexcitation of the epileptic brain. (Derived from Eiro et al. [43])

increase in AMPAR trafficking augmented the amplitude of abnormal gamma activities detected by electroencephalography (recorded at rest with eyes closure) [43]. In contrast, patients with generalized onset seizures exhibited a decrease in AMPAR coupled with increased amplitude of abnormal gamma activities [43]. Patients with epilepsy reduced cell surface AMPAR levels as compared to healthy controls, and cell surface AMPAR was reduced in larger areas of the cortex in patients with generalized onset seizures as compared to those with focal onset seizures [43]. Thus, epileptic brain function can be regulated by the enhanced trafficking of AMPAR due to Hebbian plasticity with the increased simultaneous neuronal firing and compensational downregulation of cell surface AMPAR by the synaptic scaling (Fig. 2) [43]. This indicates that two major types of synaptic plasticity, Hebbian and homeostatic synaptic plasticity, regulate epileptic brain function. Although we did not investigate surgical cases in this study, asymmetrical uptake of $[^{11}C]K$-2 colocalized with ECDs could be within the normal range due to the compensatory reduction of the physiological fraction of AMPAR in these brain areas [42].

References

1. Miyazaki T, Abe H, Uchida H, Takahashi T (2021) Translational medicine of the glutamate AMPA receptor. Proc Jpn Acad Ser B Phys Biol Sci 97:1–21. https://doi.org/10.2183/pjab. 97.001

2. Bredt DS, Nicoll RA (2003) AMPA receptor trafficking at excitatory synapses. Neuron 40: 361–379

3. Malinow R, Malenka RC (2002) AMPA receptor trafficking and synaptic plasticity. Annu Rev Neurosci 25:103–126

4. Zhu JJ, Esteban JA, Hayashi Y, Malinow R (2000) Postnatal synaptic potentiation: delivery of GluR4-containing AMPA receptors by spontaneous activity. Nat Neurosci 3:1098–1106

5. Yan D, Yamasaki M, Straub C, Watanabe M, Tomita S (2013) Homeostatic control of synaptic transmission by distinct glutamate receptors. Neuron 78:687–699. https://doi.org/ 10.1016/j.neuron.2013.02.031

6. Bliss TV, Lomo T (1973) Long-lasting potentiation of synaptic transmission in the dentate area of the anaesthetized rabbit following stimulation of the perforant path. J Physiol 232: 331–356

7. Ito M, Kano M (1982) Long-lasting depression of parallel fiber-Purkinje cell transmission induced by conjunctive stimulation of parallel fibers and climbing fibers in the cerebellar cortex. Neurosci Lett 33:253–258. https://doi. org/10.1016/0304-3940(82)90380-9

8. Huganir RL, Nicoll RA (2013) AMPARs and synaptic plasticity: the last 25 years. Neuron 80: 704–717. https://doi.org/10.1016/j.neuron. 2013.10.025

9. Takahashi T, Svoboda K, Malinow R (2003) Experience strengthening transmission by driving AMPA receptors into synapses. Science 299:1585–1588

10. Rumpel S, LeDoux J, Zador A, Malinow R (2005) Postsynaptic receptor trafficking underlying a form of associative learning. Science 308:83–88

11. Jitsuki S et al (2011) Serotonin mediates cross-modal reorganization of cortical circuits. Neuron 69:780–792

12. Mitsushima D, Ishihara K, Sano A, Kessels HW, Takahashi T (2011) Contextual learning requires synaptic AMPA receptor delivery in the hippocampus. Proc Natl Acad Sci USA 108:12503–12508., 1104558108 [pii]. https://doi.org/10.1073/pnas.1104558108

13. Mitsushima D, Sano A, Takahashi T (2013) A cholinergic trigger drives learning-induced plasticity at hippocampal synapses. Nat Commun 4:2760. https://doi.org/10.1038/ ncomms3760

14. Hayashi Y et al (2000) Driving AMPA receptors into synapses by LTP and CaMKII: requirement for GluR1 and PDZ domain interaction. Science 287:2262–2267

15. Shi S, Hayashi Y, Esteban JA, Malinow R (2001) Subunit-specific rules governing AMPA receptor trafficking to synapses in hippocampal pyramidal neurons. Cell 105:331–343

16. Diering GH, Huganir RL (2018) The AMPA receptor code of synaptic plasticity. Neuron 100:314–329. https://doi.org/10.1016/j. neuron.2018.10.018

17. Schwenk J et al (2014) Regional diversity and developmental dynamics of the AMPA-receptor proteome in the mammalian brain. Neuron 84:41–54. https://doi.org/10. 1016/j.neuron.2014.08.044

18. Shi SH et al (1999) Rapid spine delivery and redistribution of AMPA receptors after synaptic NMDA receptor activation. Science 284: 1811–1816

19. Plant K et al (2006) Transient incorporation of native GluR2-lacking AMPA receptors during hippocampal long-term potentiation. Nat Neurosci 9:602–604

20. Whitlock JR, Heynen AJ, Shuler MG, Bear MF (2006) Learning induces long-term potentiation in the hippocampus. Science 313:1093–1097

21. Takemoto K et al (2017) Optical inactivation of synaptic AMPA receptors erases fear memory. Nat Biotechnol 35:38–47. https://doi. org/10.1038/nbt.3710

22. Takemoto K et al (2011) Chromophore-assisted light inactivation of HaloTag fusion proteins labeled with eosin in living cells. ACS Chem Biol 6:401–406. https://doi.org/10. 1021/cb100431e

23. Nudo RJ (2013) Recovery after brain injury: mechanisms and principles. Front Hum Neurosci 7:887. https://doi.org/10.3389/ fnhum.2013.00887

24. Abe H et al (2018) CRMP2-binding compound, edonerpic maleate, accelerates motor function recovery from brain damage. Science 360:50–57. https://doi.org/10.1126/sci ence.aao2300

25. Goshima Y, Nakamura F, Strittmatter P, Strittmatter SM (1995) Collapsin-induced growth cone collapse mediated by an intracellular

protein related to UNC-33. Nature 376:509–514. https://doi.org/10.1038/376509a0

26. Takahashi T, Nakamura F, Jin Z, Kalb RG, Strittmatter SM (1998) Semaphorins A and E act as antagonists of neuropilin-1 and agonists of neuropilin-2 receptors. Nat Neurosci 1:487–493. https://doi.org/10.1038/2203

27. Takahashi T et al (1999) Plexin-neuropilin-1 complexes form functional semaphorin-3A receptors. Cell 99:59–69

28. Jitsuki-Takahashi A et al (2021) Activity-induced secretion of semaphorin 3A mediates learning. Eur J Neurosci 53:3279–3293. https://doi.org/10.1111/ejn.15210

29. Zhang Z et al (2020) Ketamine regulates phosphorylation of CRMP2 to mediate dendritic spine plasticity. J Mol Neurosci 70:353–364. https://doi.org/10.1007/s12031-019-01419-4

30. Ziak J et al (2020) CRMP2 mediates Sema3F-dependent axon pruning and dendritic spine remodeling. EMBO Rep 21:e48512. https://doi.org/10.15252/embr.201948512

31. Jin X et al (2016) Phosphorylation of CRMP2 by Cdk5 regulates dendritic spine development of cortical neuron in the mouse hippocampus. Neural Plast 2016:6790743. https://doi.org/10.1155/2016/6790743

32. Gu J et al (2010) ADF/cofilin-mediated actin dynamics regulate AMPA receptor trafficking during synaptic plasticity. Nat Neurosci 13:1208–1215., nn.2634 [pii]. https://doi.org/10.1038/nn.2634

33. Lee K et al (2016) AMPA receptors as therapeutic targets for neurological disorders. Adv Protein Chem Struct Biol 103:203–261. https://doi.org/10.1016/bs.apcsb.2015.10.004

34. Fu H, Chen Z, Josephson L, Li Z, Liang SH (2019) Positron emission tomography (PET) ligand development for ionotropic glutamate receptors: challenges and opportunities for radiotracer targeting N-Methyl-d-aspartate (NMDA), alpha-Amino-3-hydroxy-5-methyl-4-isoxazolepropionic Acid (AMPA), and kainate receptors. J Med Chem 62:403–419. https://doi.org/10.1021/acs.jmedchem.8b00714

35. Oi N et al (2015) Development of novel PET probes for central 2-amino-3-(3-hydroxy-5-methyl-4-isoxazolyl)propionic acid receptors. J Med Chem 58:8444–8462. https://doi.org/10.1021/acs.jmedchem.5b00712

36. Takahata K et al (2017) A human PET study of [11C]HMS011, a potential radioligand for AMPA receptors. EJNMMI Res 7:63. https://doi.org/10.1186/s13550-017-0313-0

37. Gao M, Kong D, Clearfield A, Zheng QH (2006) Synthesis of carbon-11 and fluorine-18 labeled N-acetyl-1-aryl-6,7-dimethoxy-1,2,3,4-tetrahydroisoquinoline derivatives as new potential PET AMPA receptor ligands. Bioorg Med Chem Lett 16:2229–2233. https://doi.org/10.1016/j.bmcl.2006.01.042

38. Arstad E et al (2006) Closing in on the AMPA receptor: synthesis and evaluation of 2-acetyl-1-(4'-chlorophenyl)-6-methoxy-7-[11C] methoxy-1,2,3,4-tetrahydroisoquinol ine as a potential PET tracer. Bioorg Med Chem 14:4712–4717. https://doi.org/10.1016/j.bmc.2006.03.034

39. Miyazaki T et al (2020) Visualization of AMPA receptors in living human brain with positron emission tomography. Nat Med 26:281–288. https://doi.org/10.1038/s41591-019-0723-9

40. Arisawa T et al (2021) [(11)C]K-2 image with positron emission tomography represents cell surface AMPA receptors. Neurosci Res. https://doi.org/10.1016/j.neures.2021.05.009

41. Hatano M et al (2021) Biodistribution and radiation dosimetry of the positron emission tomography probe for AMPA receptor, [(11) C]K-2, in healthy human subjects. Sci Rep 11:1598. https://doi.org/10.1038/s41598-021-81002-3

42. Miyazaki T et al (2022) Epileptic discharges initiate from brain areas with elevated accumulation of alpha-amino-3-hydroxy-5-methyl-4-isoxazole propionic acid receptors. Brain Commun 4:fcac023. https://doi.org/10.1093/braincomms/fcac023

43. Eiro T et al (2023) Dynamics of AMPA receptors regulate epileptogenesis in patients with epilepsy. Cell Rep Med 4:101020. https://doi.org/10.1016/j.xcrm.2023.101020

Chapter 3

Advanced Tau Imaging in Alzheimer's Disease and Other Types of Dementia

Kenji Tagai and Makoto Higuchi

Abstract

Following the establishment of positron emission tomography (PET) technology to visualize amyloid-β (Aβ) pathology, PET probes have been developed to capture pathological tau protein aggregates. Most clinically available probes exhibit the capability for facilitating the diagnosis of Alzheimer's disease (AD), while the detection of non-AD tau pathologies is offered by a few probes. The ultrastructure of disease-specific tau fibril folds has been revealed by cryo-electron microscopy, and the modes of the probe binding to these structures are being elucidated, potentially leading to progress in the generation of probes with high sensitivity and specificity. Tau PET holds promise for improving diagnostic accuracy, disease and subtype classifications, and staging of tauopathies. Similar to the clinical application of amyloid PET, tau PET is either approved or in clinical trials as a diagnostic tool. Additionally, tau PET is of significant utility for evaluating the impact of Aβ-targeting drugs on neuropathology and for selecting patients likely to benefit from treatment based on tau deposition levels. Furthermore, tau PET is expected to play a central role in clinical trials of therapeutics targeting tau pathologies by demonstrating mechanisms of action and providing objective outcomes.

Key words Alzheimer's disease, Frontotemporal lobar degeneration, Tau pathology, Tau PET

1 Introduction

Alzheimer's disease (AD) is pathologically characterized by the accumulation of amyloid β (Aβ) and tau lesions in the brain. Tau lesions are not only observed in AD but also in non-AD neurodegenerative diseases such as some forms of frontotemporal lobar degeneration (FTLD). Since the accumulation of tau lesions is closely related to neurodegeneration, these conditions are collectively referred to as tauopathies [1]. Visualizing tau lesions in vivo not only enables the pre-mortem diagnosis of tauopathies but also aids in understanding the pathogenesis and developing disease-modifying drugs targeting tau lesions. Among these, tau imaging using positron emission tomography (PET) and radioligands has been advancing since the early 2010s [2]. Several PET probes

Daichi Sone (ed.), *Molecular Imaging for Brain Diseases*, Neuromethods, vol. 222, https://doi.org/10.1007/978-1-0716-4494-2_3,
© The Author(s), under exclusive license to Springer Science+Business Media, LLC, part of Springer Nature 2025

capable of sensitively visualizing tau lesions have been developed, and some have been approved by the US Food and Drug Administration (FDA) for AD evaluation [3]. Various clinical trials of tau PET ligands are currently underway, and the development of disease-modifying drugs targeting tau based on tau PET imaging is also progressing. This chapter provides an overview of the evolution of tau PET probes, including clinical applications and their role in the development of disease-modifying drugs.

2   Development of Probes for Visualizing Tau Pathology

Following the development of probes visualizing amyloid-β (Aβ) aggregates using PET, PET probes for pathological tau protein aggregates have been developed by several research groups (Fig. 1). In 2005, Okamura et al. identified a series of compounds, including quinoline derivatives and benzimidazole derivatives, that bind to tau pathology and reported on their preclinical evaluation as probes [4]. The benzimidazole compound BF-128 was structurally modified by Kolb et al., leading to the creation of novel probes named T-807 and T-808 [5, 6]. Meanwhile, Higuchi et al. identified a group of chemical substances called PBB (phenyl/pyridinyl-butadienyl-benzothiazole) as ligands for various tau inclusions [7]. Clinical PET evaluations in the early 2010s demonstrated that ^{18}F-labeled T-807 (^{18}F-T807; also known as ^{18}F-AV-1451/^{18}F-flortaucipir) could detect tau deposits in Alzheimer's disease (AD) [6], and the 11C-labeled PBB compound named ^{11}C-PBB3 could visualize tau pathology in AD and frontotemporal lobar degeneration (FTLD) patient brains [7]. Quinoline

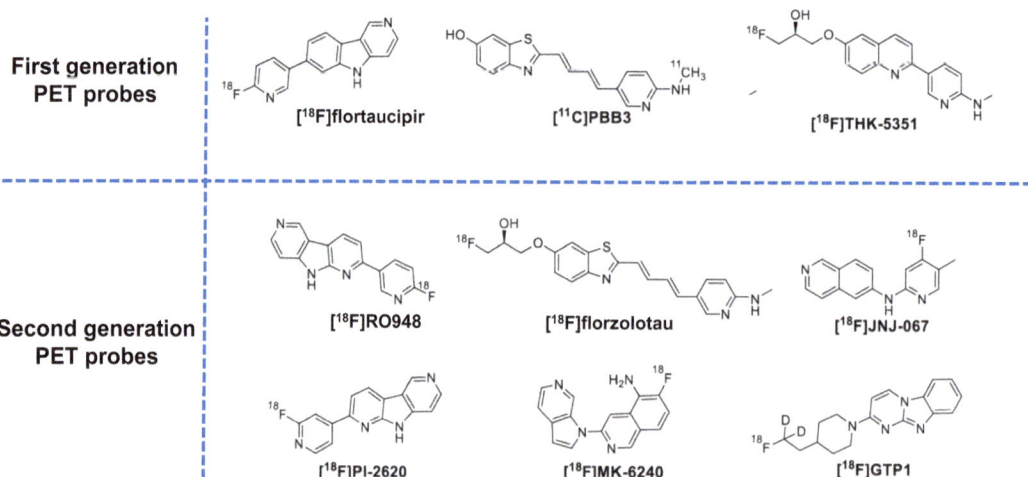

Fig. 1 The first- and second-generation tau positron emission tomography (PET) probes that have been clinically evaluated and utilized

derivatives, including ^{18}F-THK5351, were also applied to PET imaging of AD tau pathology [8]. These tau PET probes, termed first-generation tracers, have potential issues with their binding and metabolism profiles. In fact, compounds like ^{18}F-THK5351 and related quinoline compounds showed off-target binding to monoamine oxidase-B (MAO-B) [9], while 18F-flortaucipir exhibited cross-reactivity with neuromelanin-containing neurons, meningeal melanocytes, and monoamine oxidase-A (MAO-A) [10]. Although ^{11}C-PBB3 did not bind to MAO-A or MAO-B [11], it underwent rapid conversion to charged metabolites, resulting in low brain uptake compared to other tau probes [12]. Additionally, both ^{18}F-flortaucipir and ^{11}C-PBB3 showed nonspecific radioactivity accumulation in the choroid plexus [7, 13], and high radioactivity retention presumed to be from radioactive metabolites of ^{11}C-PBB3 was observed in the venous sinuses [7]. These undesirable radioactive signals may hinder the accurate evaluation of specific ligand binding in certain brain regions.

The development of second-generation tau PET probes aimed to increase signal-to-background ratios and reduce off-target binding. Most of the currently clinically available second-generation compounds are derivatives of ^{18}F-flortaucipir represented by ^{18}F-MK-6240 [14], ^{18}F-RO-9481 [15], ^{18}F-PI-2620 [16], and ^{18}F-GTP1 [17]. On the other hand, ^{18}F-PM-PBB3 (also known as ^{18}F-APN-1607/florzolotau), a fluoropropyl modification of PBB3, showed high brain penetrance with minimal metabolic susceptibility, and it did not result in significant radioactive signal accumulation in the venous sinuses, significantly surpassing ^{11}C-PBB3 in detecting tau aggregates in both AD and non-AD cases [18]. A new quinoline derivative, ^{18}F-JNJ-06717, was also evaluated in humans, but substantial off-target signals, likely attributed to MAO-B, were observed [19].

Detection of tau pathology in AD brain, along with imaging of tau deposition in primary tauopathies without Aβ deposition, can be extremely important for diagnosis and differential diagnosis of dementia and related disorders. Tau proteins in the brain are composed of six isoforms translated from selective splicing of transcripts of a single gene, which are classified into subtypes of 3-repeat (3R-tau) and 4-repeat (4R-tau) based on the number of microtubule-binding repeat domains. While AD-type tau comprises all six isoforms, in primary tauopathy such as frontotemporal lobar degeneration with tau (FTLD-tau), core lesions of diseases like progressive supranuclear palsy (PSP) and corticobasal degeneration (CBD) consist of 4R-tau isoforms, and tau inclusions named Pick bodies in Pick's disease (PiD) contain only 3R-tau isoforms [20]. Along with changes in isoform composition, there is disease-related diversity in the morphology of tau filaments visualized by electron microscopy (EM) [21]. Most second-generation probes are analogs of ^{18}F-flortaucipir, but these probes do not exhibit high

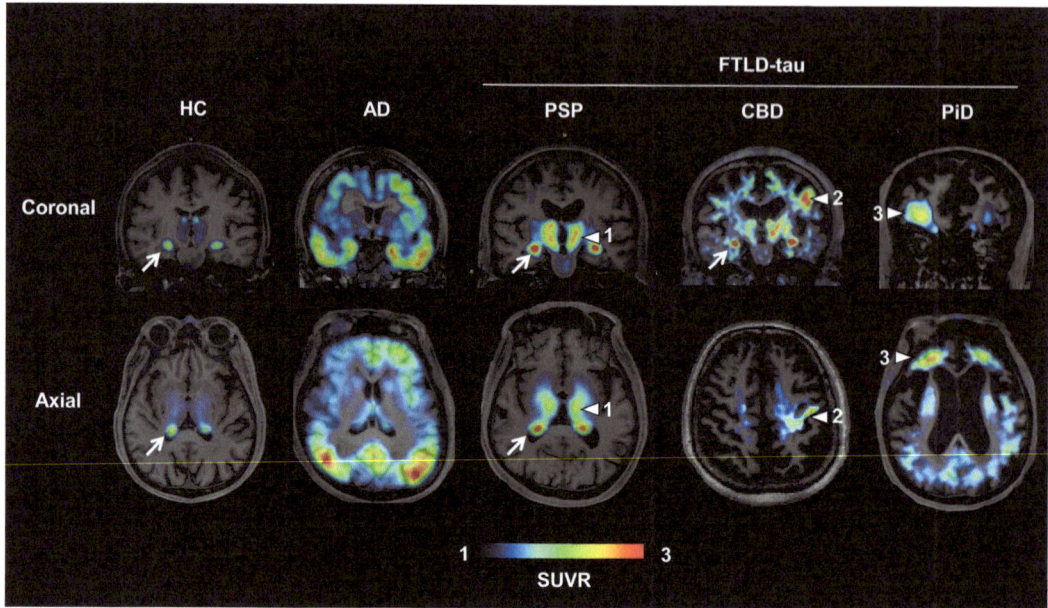

Fig. 2 PET imaging of tau lesions in the brains of AD and frontotemporal lobar degeneration (FTLD)-tau patients using ^{18}F-florzolotau. In healthy controls (HC), the probe accumulates in the TMEM106B non-tau aggregation in the choroid plexus (arrow) but shows no significant accumulation in the brain parenchyma. The choroid plexus accumulation is also observed in patients (arrow). In Alzheimer's disease (AD) patients, the probe accumulates in the limbic system and neocortex, while in progressive supranuclear palsy (PSP) patients, it accumulates in the basal ganglia, subthalamic nucleus (arrowhead 1), and brainstem. In corticobasal degeneration (CBD) patients, in addition to the regions observed in PSP, asymmetric accumulation is found in the neocortex near the primary motor cortex (arrowhead 2). In Pick's disease (PiD) patients, increased probe accumulation is seen predominantly in the anterior frontal and temporal cortex (arrowhead 3). Each FTLD-tau patient was confirmed by neuropathological assessment. (Modified from reference [18])

reactivity to non-AD-type tau aggregates [11, 22–24]. The only exception is ^{18}F-PI2620, which has been reported to accumulate in some subcortical regions of PSP patient brains [25]. On the other hand, while designed to bind to tau aggregates of all isoforms, ^{11}C-PBB3 did not provide high contrast in tau pathology of PSP or CBD due to relatively low brain uptake [26, 27]. As demonstrated by autopsy and biopsy analyses of brain tissues from subjects who underwent PET scans, ^{18}F-florzolotau can detect 3R-tau pathology in PiD and 4R-tau pathology in PSP and CBD with high sensitivity [18] (Fig. 2). Sensitivity and specificity in discriminating PSP patients, AD patients, and healthy controls exceeded 90% using standardized uptake value ratio measurements and automated analysis using machine learning algorithms [28] (Fig. 3). While ^{18}F-florzolotau may have superior lesion detection and diagnostic accuracy compared to ^{18}F-PI2620, clinical PET data comparing these two probes in the same subjects are available only for one case of FTLD-tau [29].

The AI scored from tau PET images

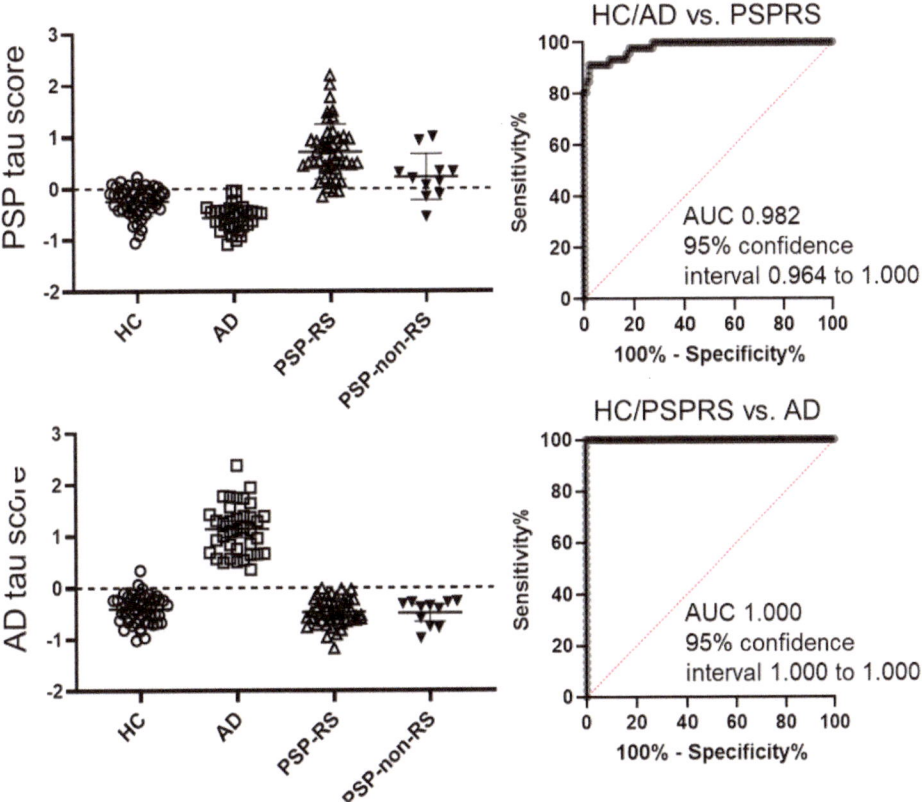

Fig. 3 Tau PET image analysis using machine learning algorithms. The artificial intelligence (AI) scored the "AD-likeness" and "PSP-likeness" of the image findings. The AD tau score and PSP tau score accurately distinguished AD and PSP patients from healthy controls (HC) and other diseases. PSP patients were analyzed separately as Richardson syndrome (PSP-RS) and non-Richardson syndrome (PSP-non-RS) patients. (Modified from Refs. [28])

Structural insights into the interaction between PET probes and AD or non-AD tau filaments are not yet sufficiently understood. Recent advancements in cryo-electron microscopy (cryo-EM) analysis have elucidated the three-dimensional structure of tau filament cores specific to individual tauopathies [30]. Furthermore, using this technique, molecules of florzolotau bound to the tau filament core in AD brains have also been visualized [31]. Tau aggregates in AD brains primarily consist of paired helical filaments (PHFs) and straight filaments, with the building blocks of these filaments formed by the dimerization of protofilaments. During this process, tau filaments form a cavity-like structure with grooved pockets, into which florzolotau binds by spanning overlapping β-sheets. The grooves are composed of peptide bond surfaces and compounds with flat and elongated skeletons that fit into the slit

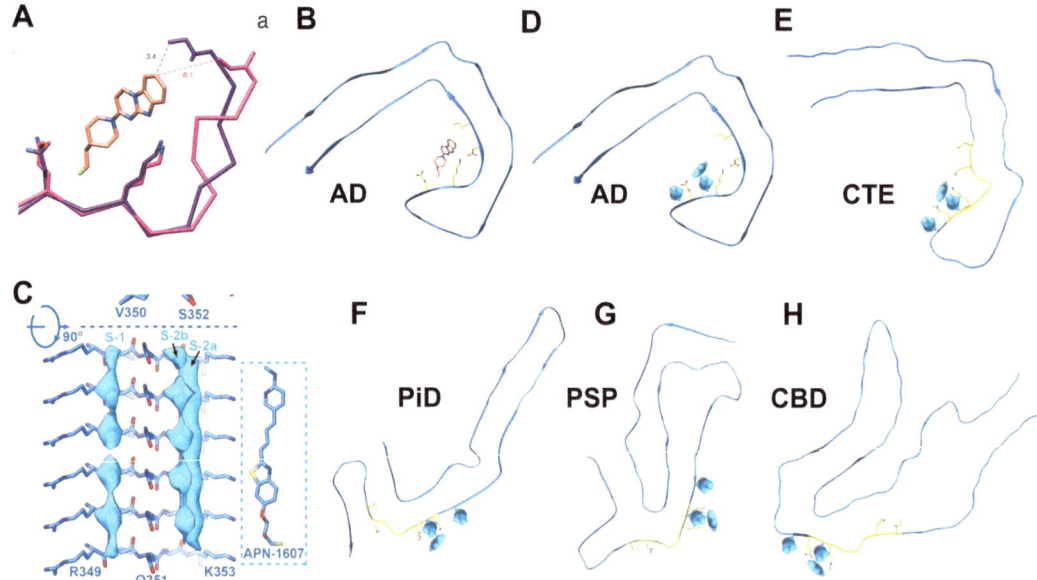

Fig. 4 Cryo-EM elucidation of PET probe binding modes to various tau fibrils. Flortaucipir analogs like GTP-1 and MK-6240 bind to a pocket formed by the side chains of four amino acid residues in AD tau fibrils (**a**, **b**), which is absent in non-AD tau fibrils (**e–h**). Conversely, florzolotau binds to a groove-like pocket formed by the stacking of tau protofibrils (**c**, **d**). This groove-like pocket is present in various types of tau fibrils, allowing florzolotau to bind to multiple fibril types (**e–h**, compound shown in blue perpendicular to the plane of the figure). (Modified from Refs. [31, 32])

bind. Some grooves, formed by hydrophilic side chains, significantly open up due to interaction with adjacent residue side chains, making them more prone to hydrophobic interactions with florzolotau [21, 31]. Additionally, further binding of florzolotau occurs in parallel, stacking on florzolotau bound to the groove.

More recently, cryo-EM analysis has revealed the binding modes of flortaucipir-like probes, including GTP-1 and MK-6240, with PHFs derived from AD brains (Fig. 4). Within PHFs, side chains of four amino acid residues form pockets, into which GTP-1 and MK-6240 bind snugly [32, 33]. On the other hand, in non-AD-type 4-repeat tau or 3-repeat tau fibers, the four side chains do not form pockets, making binding less likely. However, since the aforementioned hydrophobic grooves are present in all tau aggregates, florzolotau can bind to various tau filaments. Furthermore, while it is known that tau fibers in chronic traumatic encephalopathy (CTE) are composed of all six isoforms, there are no highly binding pockets for flortaucipir-type probes. Conversely, the hydrophobic grooves are more exposed to AD-type fibers, making them more accessible for florzolotau binding. Indeed, in actual clinical PET evaluations, ^{18}F-flortaucipir and ^{18}F-MK-6240 have shown limited utility in detecting tau deposits in CTE cases [34, 35]. However, exploratory PET studies using ^{18}F-florzolotau have demonstrated its ability to capture CTE tau lesions with sufficiently high contrast [36]. Furthermore, it has recently been revealed by cryo-EM that CTE-type tau fibers constitute the core pathology of subacute sclerosing panencephalitis [37] and Kii ALS/PDC [38], suggesting that CTE-type tau aggregates are formed in association with protracted neuroinflammation and can be detected with high sensitivity using florzolotau-type PET probes.

3 Diagnosis and Prognosis Prediction with Tau PET

In the spectrum from normal aging to AD, tau deposits can be tracked by PET as they spread from the entorhinal cortex to the limbic system and subsequently to the neocortex, following the neuropathological stages of tau proposed by Braak [39, 40] (Fig. 2). Unlike amyloid PET probe brain accumulation, which reaches a ceiling in the mild cognitive impairment (MCI) stage, detectable tau accumulation via PET varies throughout the entire AD spectrum, thus aiding in objectively assessing disease severity from early to late stages [41, 42]. Clinical findings from tau PET after the onset of AD provide significantly higher specificity in AD diagnosis compared to other modalities like MRI [43, 44]. Based on the demonstrated utility in clinical trials, ^{18}F-flortaucipir has been approved by the US Food and Drug Administration for AD diagnosis, with clinical trials ongoing for ^{18}F-florzolotau and ^{18}F-MK-6240 targeting AD and MCI. Longitudinal PET scans have revealed the progression of AD tau pathology within the same individual, which can serve as reference data for conducting clinical trials of therapeutic agents [45]. Additionally, data-driven approaches have shown diversity in tau distribution in AD brains, classifying several subtypes based on the predominance of lesions in the inner part of the temporal lobe or the posterior or lateral parts of the temporal lobe [46]. By combining unbiased subtype classification with conventional region-based quantification methods, PET-based stratification of AD patients is expected, allowing for the prediction of prognosis and treatment effects. Potential issues in the unbiased evaluation of AD tau pathology include differences in contrast and nonspecific accumulation among different probes. Comparative image analysis proposes the CenTauRz method for standardizing and normalizing probe accumulation data obtained from PET measurements using ^{18}F-flortaucipir, ^{18}F-MK-6240, ^{18}F-PI-2620, ^{18}F-florzolotau, ^{18}F-GTP-1, and ^{18}F-RO-948 in AD spectrum patients [47].

While the utility of tau PET for diagnosis and progression prediction in the early stages of AD has been demonstrated, the significance of tau PET for subjects with MCI and those without cognitive decline requires further rigorous analysis.

Image-pathology correlation studies have shown that positive findings in ^{18}F-flortaucipir-PET images in dementia patients are closely related to Braak stages V/VI tau pathology [48], justifying the use of this probe's PET scans for neuropathological evaluation of dementia patient brains in the clinic. Conversely, the detection accuracy of Braak stage IV or earlier tau pathology with ^{18}F-flortaucipir in individuals without cognitive impairment or in MCI cases was insufficient. This issue may stem from variability in tau filament maturity and density at these stages, as well as the difficulty in accurately assessing probe-specific binding to tau deposits in the hippocampus due to spillover from the choroid plexus. Similarly, region-of-interest-based semi-quantitative assessments showed lower diagnostic accuracy for MCI due to AD compared to AD dementia [42–44]. Furthermore, tau pathology below Braak stage V may manifest without significant amyloid accumulation, known as primary aging-related tauopathy (PART), which can be visualized by tau PET [49, 50]. While technical challenges remain regarding diagnostic utility in preclinical AD and AD-related MCI, tau PET has been demonstrated to be highly useful in predicting cognitive decline in these patients compared to amyloid PET [51]. Indeed, among cognitively unimpaired older adults, individuals positive for both amyloid PET and tau PET had a 6- to 8-fold higher risk of progressing to MCI compared to those positive only for amyloid PET.

4   Clinical Applications and Drug Development Based on Tau PET

Similar to amyloid PET, tau PET is increasingly being introduced into clinical practice and disease-modifying therapy development. Research supporting the idea that tau PET enhances diagnostic reliability and significantly influences decision-making processes for therapeutic interventions has been conducted [52, 53]. Particularly, tau PET is employed as an outcome measure in clinical trials of anti-amyloid therapies, and treatment with aducanumab and lecanemab has been shown to delay the progression of tau pathology detected by PET [54, 55]. However, in phase III trials of donanemab, no significant inhibition of tau accumulation measured by PET was demonstrated [56]. Despite this, discrepancies seem to exist between clinical evaluations and tau PET assessments, as cognitive decline was inhibited by 30–35% with lecanemab and donanemab administration. Donanemab targets a variant of Aβ called AβN3pE, hypothesized to primarily mitigate the neurotoxicity of this variant to slow cognitive impairment progression. In contrast, lecanemab targets Aβ oligomers, presumed to mitigate the complex toxicity of oligomers and resulting tau aggregates. However, these hypotheses have not been sufficiently validated at this point. Although donanemab did not alter tau PET findings, significant inhibition of clinical progression with donanemab was achieved by selectively treating patients with mild to moderate tau deposition on pretreatment tau PET scans [56].

Moreover, in clinical trials of therapeutics targeting tau pathology itself, tau PET has begun to provide mechanistic evidence and outcome measures [57, 58]. The utility of tau imaging has been demonstrated in phase Ib clinical trials of an antisense oligonucleotide (ASO) called BIIB080, which suppresses tau gene expression in early AD patients. Remarkably, tau PET probe accumulation in the temporal lobe and its adjacent regions decreased after one year of BIIB080 treatment [58, 59]. This suggests bidirectional conversion between tau monomers in vivo and detectable high molecular weight tau fibrils by PET, indicating that tau aggregates can be degraded and removed by homeostatic mechanisms. However, whether the reduction in tau PET signal induced by treatment is associated with improvements in clinical symptoms remains unclear. It is anticipated that as clinical trials progress for primary tauopathies such as PSP, ASOs, and other anti-tau therapeutics, the evaluation of 4-repeat or 3-repeat tau deposition with PET probes like ^{18}F-florzolotau will become crucial for appropriate patient selection and efficacy validation.

References

1. Iqbal K, Liu F, Gong C-X (2016) Tau and neurodegenerative disease: the story so far. Nat Rev Neurol 12:15–27

2. Villemagne VL, Doré V, Burnham SC et al (2018) Imaging tau and amyloid-β proteinopathies in Alzheimer disease and other conditions. Nat Rev Neurol 14:225–236

3. Jie CVML, Treyer V, Schibli R, Mu L (2021) Tauvid^{TM}: the first FDA-approved PET tracer for imaging tau pathology in Alzheimer's disease. Pharmaceuticals 14. https://doi.org/10.3390/ph14020110

4. Okamura N, Suemoto T, Furumoto S et al (2005) Quinoline and benzimidazole derivatives: candidate probes for in vivo imaging of tau pathology in Alzheimer's disease. J Neurosci 25:10857–10862

5. Chien DT, Szardenings AK, Bahri S et al (2014) Early clinical PET imaging results with the novel PHF-tau radioligand [F18]-T808. J Alzheimers Dis 38:171–184

6. Chien DT, Bahri S, Szardenings AK et al (2013) Early clinical PET imaging results with the novel PHF-tau radioligand [F-18]-T807. J Alzheimers Dis 34:457–468

7. Maruyama M, Shimada H, Suhara T et al (2013) Imaging of tau pathology in a tauopathy mouse model and in Alzheimer patients compared to normal controls. Neuron 79:1094–1108

8. Okamura N, Furumoto S, Fodero-Tavoletti MT et al (2014) Non-invasive assessment of Alzheimer's disease neurofibrillary pathology using 18F-THK5105 PET. Brain 137:1762–1771

9. Ng KP, Pascoal TA, Mathotaarachchi S et al (2017) Monoamine oxidase B inhibitor, selegiline, reduces 18F-THK5351 uptake in the human brain. Alzheimers Res Ther 9:25

10. Marquié M, Normandin MD, Vanderburg CR et al (2015) Validating novel tau positron emission tomography tracer [F-18]-AV-1451 (T807) on postmortem brain tissue. Ann Neurol 78:787–800

11. Ono M, Sahara N, Kumata K et al (2017) Distinct binding of PET ligands PBB3 and AV-1451 to tau fibril strains in neurodegenerative tauopathies. Brain 140:764–780

12. Hashimoto H, Kawamura K, Igarashi N et al (2014) Radiosynthesis, photoisomerization, biodistribution, and metabolite analysis of 11C-PBB3 as a clinically useful PET probe for imaging of tau pathology. J Nucl Med 55:1532–1538

13. Ikonomovic MD, Abrahamson EE, Price JC et al (2016) [F-18]AV-1451 positron emission

tomography retention in choroid plexus: More than "off-target" binding. Ann Neurol 80: 307–308

14. Aguero C, Dhaynaut M, Normandin MD et al (2019) Autoradiography validation of novel tau PET tracer [F-18]-MK-6240 on human postmortem brain tissue. Acta Neuropathol Commun 7:37

15. Honer M, Gobbi L, Knust H et al (2018) Preclinical evaluation of 18F-RO6958948, 11C-RO6931643, and 11C-RO6924963 as novel PET radiotracers for imaging tau aggregates in Alzheimer disease. J Nucl Med 59: 675–681

16. Kroth H, Oden F, Molette J et al (2019) Discovery and preclinical characterization of [18F] PI-2620, a next-generation tau PET tracer for the assessment of tau pathology in Alzheimer's disease and other tauopathies. Eur J Nucl Med Mol Imaging 46:2178–2189

17. Sanabria Bohórquez S, Marik J, Ogasawara A et al (2019) [18F]GTP1 (Genentech Tau Probe 1), a radioligand for detecting neurofibrillary tangle tau pathology in Alzheimer's disease. Eur J Nucl Med Mol Imaging 46: 2077–2089

18. Tagai K, Ono M, Kubota M et al (2021) High-contrast in vivo imaging of tau pathologies in Alzheimer's and non-Alzheimer's disease tauopathies. Neuron 109:42–58.e8

19. Baker SL, Provost K, Thomas W et al (2021) Evaluation of [18F]-JNJ-64326067-AAA tau PET tracer in humans. J Cereb Blood Flow Metab 41:3302–3313

20. Wang Y, Mandelkow E (2016) Tau in physiology and pathology. Nat Rev Neurosci 17:5–21

21. Sahara N, Higuchi M (2024) Diagnostic and therapeutic targeting of pathological tau proteins in neurodegenerative disorders. FEBS Open Bio 14:165–180

22. Leuzy A, Chiotis K, Lemoine L et al (2019) Tau PET imaging in neurodegenerative tauopathies-still a challenge. Mol Psychiatry 24:1112–1134

23. Marquié M, Normandin MD, Meltzer AC et al (2017) Pathological correlations of [F-18]-AV-1451 imaging in non-Alzheimer tauopathies. Ann Neurol 81:117–128

24. Malarte M-L, Gillberg P-G, Kumar A et al (2023) Discriminative binding of tau PET tracers PI2620, MK6240 and RO948 in Alzheimer's disease, corticobasal degeneration and progressive supranuclear palsy brains. Mol Psychiatry 28:1272–1283

25. Brendel M, Barthel H, van Eimeren T et al (2020) Assessment of 18F-PI-2620 as a biomarker in progressive supranuclear palsy. JAMA Neurol 77:1408–1419

26. Nakano Y, Shimada H, Shinotoh H et al (2022) PET-based classification of corticobasal syndrome. Parkinsonism Relat Disord 98:92–98

27. Endo H, Shimada H, Sahara N et al (2019) In vivo binding of a tau imaging probe, [11 C] PBB3, in patients with progressive supranuclear palsy. Mov Disord 34:744–754

28. Endo H, Tagai K, Ono M et al (2022) A machine learning-based approach to discrimination of tauopathies using [18 F]PM-PBB3 PET images. Mov Disord 37:2236–2246

29. Tezuka T, Takahata K, Seki M et al (2021) Evaluation of [18F]PI-2620, a second-generation selective tau tracer, for assessing four-repeat tauopathies. Brain Commun 3: fcab190

30. Shi Y, Zhang W, Yang Y et al (2021) Structure-based classification of tauopathies. Nature 598: 359–363

31. Shi Y, Murzin AG, Falcon B et al (2021) Cryo-EM structures of tau filaments from Alzheimer's disease with PET ligand APN-1607. Acta Neuropathol 141:697–708

32. Merz GE, Chalkley MJ, Tan SK et al (2023) Stacked binding of a PET ligand to Alzheimer's tau paired helical filaments. Nat Commun 14: 3048

33. Kunach P, Vaquer-Alicea J, Smith MS et al (2023) Cryo-EM structure of Alzheimer's disease tau filaments with PET ligand MK-6240. bioRxivorg. https://doi.org/10.1101/2023. 09.22.558671

34. Alosco ML, Iaccarino L, Asken BM et al (2022) 18F-MK-6240 tau PET as a biomarker for chronic traumatic encephalopathy: case series of 10 symptomatic former national football league players. Alzheimers Dement 18. https://doi.org/10.1002/alz.066995

35. Alosco ML, Su Y, Stein TD et al (2023) Associations between near end-of-life flortaucipir PET and postmortem CTE-related tau neuropathology in six former American football players. Eur J Nucl Med Mol Imaging 50: 435–452

36. Takahata K, Shimada H, Kubota M et al (2020) Grouping tau topologies in former boxers assessed by PET with 18 F-PM-PBB3: a reappraisal of dementia pugilistica. Alzheimers Dement 16. https://doi.org/10.1002/alz. 041510

37. Qi C, Hasegawa M, Takao M et al (2023) Identical tau filaments in subacute sclerosing panencephalitis and chronic traumatic encephalopathy. Acta Neuropathol Commun 11:74

38. Qi C, Verheijen BM, Kokubo Y et al (2023) Tau filaments from amyotrophic lateral sclerosis/parkinsonism-dementia complex (ALS/PDC) adopt the CTE fold. bioRxivorg. https://doi.org/10.1101/2023.04.26.538417

39. Braak H, Alafuzoff I, Arzberger T et al (2006) Staging of Alzheimer disease-associated neurofibrillary pathology using paraffin sections and immunocytochemistry. Acta Neuropathol 112:389–404

40. Braak H, Braak E (1995) Staging of Alzheimer's disease-related neurofibrillary changes. Neurobiol Aging 16:271–278. discussion 278–84

41. Schöll M, Lockhart SN, Schonhaut DR et al (2016) PET imaging of tau deposition in the aging human brain. Neuron 89:971–982

42. Pascoal TA, Therriault J, Benedet AL et al (2020) 18F-MK-6240 PET for early and late detection of neurofibrillary tangles. Brain 143:2818–2830

43. Leuzy A, Smith R, Ossenkoppele R et al (2020) Diagnostic performance of RO948 F 18 tau positron emission tomography in the differentiation of alzheimer disease from other neurodegenerative disorders. JAMA Neurol 77:955–965

44. Ossenkoppele R, Rabinovici GD, Smith R et al (2018) Discriminative accuracy of [18F]flortaucipir positron emission tomography for alzheimer disease vs other neurodegenerative disorders. JAMA 320:1151–1162

45. Leuzy A, Binette AP, Vogel JW et al (2023) Comparison of group-level and individualized brain regions for measuring change in longitudinal tau positron emission tomography in Alzheimer disease. JAMA Neurol. https://doi.org/10.1001/jamaneurol.2023.1067

46. Vogel JW, Young AL, Oxtoby NP et al (2021) Four distinct trajectories of tau deposition identified in Alzheimer's disease. Nat Med 27:871–881

47. Villemagne VL, Leuzy A, Bohorquez SS et al (2023) CenTauR: toward a universal scale and masks for standardizing tau imaging studies. Alzheimers Dement 15:e12454

48. Fleisher AS, Pontecorvo MJ, Devous MD Sr et al (2020) Positron emission tomography imaging with [18F]flortaucipir and postmortem assessment of Alzheimer disease neuropathologic changes. JAMA Neurol 77:829–839

49. Wuestefeld A, Pichet Binette A, Berron D et al (2023) Age-related and amyloid-beta-independent tau deposition and its downstream effects. Brain. https://doi.org/10.1093/brain/awad135

50. Villemagne VL, Lopresti BJ, Doré V et al (2021) What is T+? A gordian knot of tracers, thresholds, and topographies. J Nucl Med 62:614–619

51. Ossenkoppele R, Smith R, Mattsson-Carlgren N et al (2021) Accuracy of tau positron emission tomography as a prognostic marker in preclinical and prodromal Alzheimer disease: a head-to-head comparison against amyloid positron emission tomography and magnetic resonance imaging. JAMA Neurol 78:961–971

52. Shimohama S, Tezuka T, Takahata K et al (2023) Impact of amyloid and tau PET on changes in diagnosis and patient management. Neurology 100:e264–e274

53. Smith R, Hägerström D, Pawlik D et al (2023) Clinical utility of tau positron emission tomography in the diagnostic workup of patients with cognitive symptoms. JAMA Neurol. https://doi.org/10.1001/jamaneurol.2023.1323

54. Budd Haeberlein S, Aisen PS, Barkhof F et al (2022) Two randomized phase 3 studies of aducanumab in early Alzheimer's disease. J Prev Alzheimers Dis 9:197–210

55. van Dyck CH, Swanson CJ, Aisen P et al (2023) Lecanemab in early Alzheimer's disease. N Engl J Med 388:9–21

56. Sims JR, Zimmer JA, Evans CD et al (2023) Donanemab in early symptomatic Alzheimer disease: the TRAILBLAZER-ALZ 2 randomized clinical trial. JAMA 330:512–527

57. Teng E, Manser PT, Pickthorn K et al (2022) Safety and efficacy of semorinemab in individuals with prodromal to mild Alzheimer disease: a randomized clinical trial. JAMA Neurol 79:758–767

58. Edwards AL, Collins JA, Junge C et al (2023) Exploratory tau biomarker results from a multiple ascending-dose study of BIIB080 in Alzheimer disease: a randomized clinical trial. JAMA Neurol 80:1344–1352

59. Mummery CJ, Börjesson-Hanson A, Blackburn DJ et al (2023) Tau-targeting antisense oligonucleotide MAPTRx in mild Alzheimer's disease: a phase 1b, randomized, placebo-controlled trial. Nat Med 29:1437–1447

Chapter 4

PET Visualization of Brain Tau Accumulations Secondary to Various CNS Injuries: Chronic Traumatic Encephalopathy (CTE) and Organophosphorus Poisoning

Keisuke Takahata, Sho Moriguchi, Hisaomi Suzuki, Shin Kurose, Yuki Momota, Kenji Tagai, Hironobu Endo, Yuko Kataoka, Masanori Ichihashi, Yuki Komatsu, Sachiko Anamizu, Naruhiko Sahara, and Makoto Higuchi

Abstract

Traumatic brain injury (TBI) can lead to severe and persistent neuropsychiatric sequelae. Recent clinical and neuropathological studies have elucidated the epidemiological link between TBI and the subsequent development of neurodegenerative diseases. Chronic traumatic encephalopathy (CTE), a neurodegenerative disorder associated with repeated mild TBI, is characterized by the deposition of hyperphosphorylated tau in neurofibrillary and astrocytic tangles, predominantly surrounding small blood vessels in the cortical sulci. Recent cryo-electron microscopy (cryo-EM) studies have shown that CTE-type tau folds appear in the brain long after the onset of some central nervous system diseases. In addition to TBI-related tauopathies, emerging evidence suggests that exposure to organophosphorus compounds, such as pesticides and nerve agents, may also contribute to long-term neuropsychiatric sequelae, including cognitive impairment and psychiatric disorders, and has been postulated to promote neuropathological changes including tau accumulation through persistent neuroinflammation and excitotoxicity. Positron emission tomography (PET) has emerged as a valuable tool for detecting molecular pathologies in neurodegenerative diseases. This review focuses on the development of tau PET tracers for in vivo visualization of tau deposition in chronic TBI states. We present current findings from tau PET imaging studies on CTE and other secondary tauopathies, utilizing both first- and second-generation tau PET tracers. Additionally, we discuss PET quantification techniques for assessing tau burden in the chronic stages of TBI, with particular emphasis on the unique topography of tau lesions in CTE.

Key words Chronic traumatic encephalopathy, Traumatic brain injury, Concussion, Tau, Secondary tauopathy, Positron emission tomography, Biomarker, Florzolotau, Organophosphorus

1 Introduction: Chronic Traumatic Encephalopathy

Recent neuropathological studies have shown that a variety of neurodegenerative diseases can be caused long after traumatic brain injury (TBI). A typical delayed-onset condition caused by

Daichi Sone (ed.), *Molecular Imaging for Brain Diseases*, Neuromethods, vol. 222, https://doi.org/10.1007/978-1-0716-4494-2_4,
© The Author(s), under exclusive license to Springer Science+Business Media, LLC, part of Springer Nature 2025

mild repetitive TBI is chronic traumatic encephalopathy (CTE) [1]. Historically, long-term neuropsychiatric sequelae resulting from mild repetitive TBI were considered a rare condition, primarily affecting retired boxers and other combat sports athletes. However, recent reports indicate that a wide range of contact sports and occupations may contribute to the development of CTE [1, 2]. Furthermore, recent research has revealed that CTE-type tau fibers accumulate due to various central nervous system (CNS) injuries beyond TBI [3]. This finding expands our understanding of the condition's etiology and potential risk factors.

CTE represents a significant public health concern, as it can lead to early-onset dementia, severe depressive symptoms, and behavioral disturbances. Currently, CTE can only be definitively diagnosed through neuropathological assessment of postmortem brain tissue, highlighting the critical need for reliable biomarkers to detect neuropathological changes associated with CTE in living individuals. Given these challenges, current research efforts in this field are focused on developing reliable biomarkers to detect neuropathological changes of CTE in living subjects. This approach aims to facilitate early diagnosis, monitor disease progression, and potentially guide therapeutic interventions.

1.1 Association of CTE and Repetitive Mild TBI

The etiology of CTE has been significantly expanded, with evidence now confirming its occurrence in former athletes across a wide range of contact sports, including American football [4], rugby [5], soccer [6], wrestling [7], ice hockey [8], baseball [9], and boxing [10]. While CTE has also been observed in veterans exposed to blast injuries, its prevalence appears to be lower in this population compared to contact sports athletes [11]. Interestingly, postmortem studies have revealed neuropathological changes similar to CTE in individuals with refractory seizures [12], self-injurious behavior associated with developmental disorders [13], and those exposed to domestic violence [14]. These findings underscore that CTE is not a rare condition, but rather a potential consequence of various central nervous system (CNS) injuries involving repetitive mild TBI, other types of CNS injuries.

A recent survey has revealed that the prevalence of CTE in populations with contact sports experience was extremely high. A notable study from Boston University reported that 87% of postmortem brains from individuals with a history of college-level American football exhibited tau accumulation consistent with CTE neuropathology. More strikingly, 99% of former professional NFL players showed CTE-associated tau accumulations [4], indicating an extraordinarily high frequency of CTE pathology following prolonged exposure to repetitive mild TBI.

CTE can manifest at a young age, with the youngest confirmed case reported in a 17-year-old patient [1]. Neuropathological assessments of individuals who died young have revealed that the neuropathological and clinical features of young-onset CTE differ from those developing later in middle age. A comprehensive neuropathological study of 152 contact sports athletes who died before the age of 30 found that 41.4% exhibited CTE pathology [15]. Although 95.2% of these cases were in the early stages of CTE (Stages I and II), these findings suggest that tauopathy can emerge early after exposure to contact sports, affecting nearly half of the young athletes in American football.

1.2 Clinical Symptoms of CTE

Cognitive impairment stands as the most prevalent neuropsychiatric manifestation of CTE, with memory deficits being the most frequently observed symptom. Impairments in working memory and attention are also commonly reported [16]. Furthermore, CTE patients often exhibit aggression and violent behavior, likely stemming from difficulties in emotional regulation. Headache is a common physical symptom associated with the condition.

The progression of CTE symptoms can vary significantly between individuals. While some cases demonstrate a slow, gradual progression, others may remain at a particular stage without obvious clinical deterioration [17]. Notably, CTE is characterized by a higher frequency of psychiatric symptoms, including depression and delusions, and is associated with an elevated risk of suicide [18, 19]. Additionally, there is a higher prevalence of substance abuse disorders, including alcoholism and drug abuse, as well as gambling addiction among CTE patients. These behavioral issues strongly suggest the presence of inhibitory dysfunction resulting from TBI. A clear correlation exists between the degree of pathological progression in CTE and the severity of clinical symptoms [1]. As the neuropathological stage increases, cognitive decline, psychiatric symptoms, and motor symptoms progressively increase, leading to dementia in almost all patients at the most severe stage, Stage IV [1].

1.3 Neuropathological Features of CTE

CTE is a distinct neurodegenerative disease characterized by the deposition of hyperphosphorylated tau in the form of neurofibrillary and astrocytic tangles. These deposits are typically found around small blood vessels of the cortex, predominantly at sulcal depths [1] (Fig. 1). Biochemical analysis of tau isoforms in CTE has revealed the presence of both three-repeat and four-repeat tau, similar to the pattern observed in Alzheimer's disease (AD) [1]. The four-repeat isoform of tau is predominantly expressed in astrocytes located in the subpial region of sulcal depths [1, 16], although some neuronal abnormalities in the hippocampal region also appear to be primarily four-repat tau in CTE.

A distinctive feature of CTE is the irregular and patchy distribution of tau lesions, which contrasts with the more uniform distribution observed in other neurodegenerative diseases such as AD. Consistent neuropathological features of CTE include axonal

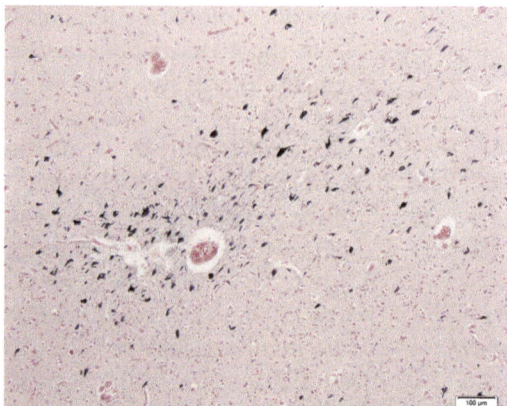

Fig. 1 Neuropathology of chronic traumatic encephalopathy (CTE) The typical neuropathology of chronic traumatic encephalopathy. Gallyas-Braak silver staining was used to visualize the neurofibrillary tangles around the small vessels at the depth of the cortical sulcus

injury, axonal degeneration, myelin degeneration, and white-matter loss, which may play a critical role in initiating p-tau pathology.

In addition to tau lesions, CTE is characterized by a high prevalence of TAR DNA-binding protein-43 kDa (TDP-43)-positive inclusions [20]. These ubiquitin-positive inclusions are also found in frontotemporal lobar degeneration (FTLD) and amyotrophic lateral sclerosis (ALS). In advanced cases of CTE, accumulations of α-synuclein and β-amyloid are also observed in the brain [1]. These findings indicate that CTE is a neurodegenerative disease with multiple pathologies of protein accumulation, or polyproteinopathy. The mechanisms underlying the aggregation and propagation of tau and other proteins, and how these processes lead to changes in brain circuitry, remain unclear. This gap in our understanding highlights the need for further research into the molecular and cellular pathways involved in CTE pathogenesis.

1.4 Tauopathy Following Moderate-to-Severe TBI

While the acute clinical manifestations of severe TBI differ from those of mild-repetitive TBI, their long-term neuropsychiatric consequences share certain features [21]. Progressive cognitive decline has been observed in the chronic stage of severe TBI, often accelerating age-related cognitive deterioration [22]. Psychiatric symptoms, particularly difficulties in emotion regulation, are distinct features frequently observed in long-term severe TBI survivors.

In some cases, schizophrenia-like psychotic states emerge late after a single incidence of moderate-to-severe TBI. Psychosis that appear later after severe TBI have been described as psychotic disorders following TBI (PDFTBI) [23–25]. Studies indicate that PDFTBI occurs in 3.4–8.9% of severe TBI cases, with an average time from injury to symptom onset of approximately 4 to 6 years [24, 26]. PDFTBI is typically associated with damage to the temporal pole and frontal lobe but can also result from damage to other brain areas. In some cases, widespread white matter lesions due to diffuse axonal injury can cause late-onset psychosis, even in the absence of focal brain damage. Most cases exhibit impairments in executive function and memory, as well as attentional deficits, in addition to psychotic symptoms such as delusions and hallucinations. PDFTBI shares with chronic traumatic encephalopathy in that it develops long after the TBI [1, 27]. The long latency between head injury and the emergence of mental disorders strongly suggests that neurodegenerative processes induced by severe TBI play a role in the development of psychotic symptoms [27].

Amyloid β (Aβ) pathologies have been extensively studied in relation to dementia following severe TBI. Early observations revealed that 30% of patients who died in the acute phase of moderate to severe TBI had Aβ plaques and more pronounced Aβ pathological changes than age-matched controls [28–30]. While less frequently reported, tau accumulation has been observed in the brains of severe TBI cases [31]. A study of 39 individuals who survived moderate to severe TBI for more than 1 year showed a higher density and more widespread distribution of neurofibrillary tangles (NFTs) compared to age-matched controls. These NFTs, found at the depth of cortical sulci, exhibited features similar to those seen in CTE. Tau lesions similar to CTE were found in the postmortem brains of two patients who had undergone prefrontal leucotomy [32]. Neuropathological findings of CTE have also been confirmed in cases of brain damage due to blast injury [33].

Currently, it remains inconclusive whether CTE is specific to mild-repetitive TBI. Evidence suggests that chronic pathologies following severe and mild-repetitive injuries show similarities [34], although comparative studies are limited. This highlights the need for further research to elucidate the relationship between different types of TBI and their long-term neuropathological consequences.

1.5 Secondary Tauopathy Following Various CNS Diseases: CTE-Type Pathology

Emerging evidence from postmortem research suggests that tau accumulation is not exclusive to TBI but can be induced by a variety of CNS injuries. This expanding body of knowledge has led to the concept of CTE-type pathology occurring in diverse neurological conditions.

Postmortem brain studies have revealed neurofibrillary tangles in the brains of patients with subacute sclerosing panencephalitis (SSPE), a rare complication of measles infection [35]. Notably, a 2023 cryo-electron microscopy (cryo-EM) study demonstrated that the three-dimensional structure of tau folds in SSPE was identical to that observed in Chronic Traumatic Encephalopathy

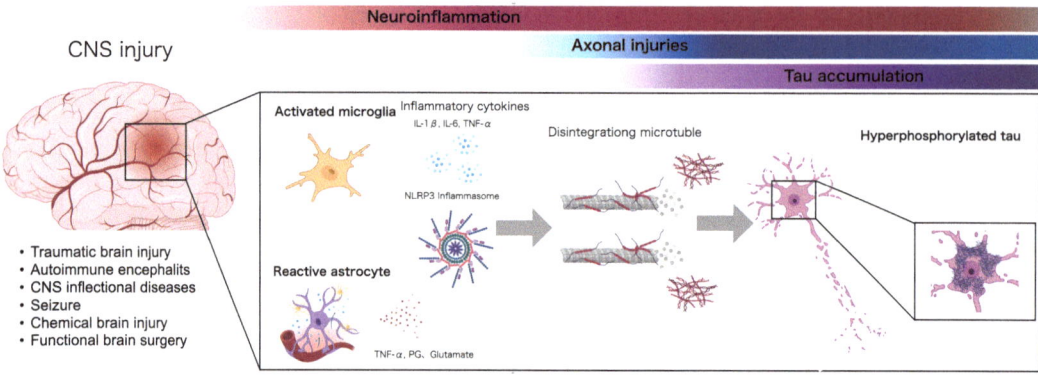

Fig. 2 Secondary tauopathies following various central nervous system (CNS) diseases. Traumatic brain injury, autoimmune encephalitis, CNS infectious diseases, epilepsy, toxicosis, and environmental factors have been implicated as causes of secondary tauopathy. Persistent neuroinflammation and excitotoxicity are presumed to be factors that accelerate tau accumulation

(CTE) [36]. Similarly, patients with Amyotrophic Lateral Sclerosis/Parkinsonism-Dementia Complex (ALS/PDC), an endemic disease in the Kii Peninsula of Japan and Guam, were found to accumulate tau fibers with CTE-type folding [37]. This finding suggests a potential link between environmental factors and the development of CTE-like tauopathy.

In addition, tau accumulation in the brain is reportedly caused by brain lobe dissection [32], refractory epilepsy [12], heavy metal poisoning [38], Nodding syndrome (a form of epilepsy locally recognized in Sudan and other African countries) [39], and autoimmune-mediated encephalitis (anti-LGI1 antibody encephalitis and anti-IgLON5 antibody encephalitis) [3, 40, 41]. These reports on tau deposition following various CNS diseases suggest that tauopathy may be a common late-onset condition in a wide range of CNS injuries. The underlying mechanisms driving these secondary tauopathies are not fully understood, but persistent neuroinflammation has been postulated as a possible cause in various CNS injuries [3] (Fig. 2).

1.6 Neuro-pathological Sequelae of Organophosphorus Poisoning

Organophosphorus is a chemical compound class containing phosphorus atoms bonded to carbon atoms. Organophosphorus has been used as a pesticide, and compounds with particularly strong neurotoxicity in humans have been used as chemical weapons. Exposure to organophosphorus has been associated with long-term neurological sequelae, including a condition known as Chronic Organophosphate-Induced Neuropsychiatric Disorder (COPIND). COPIND typically manifests months to years after exposure and persists over time [42]. The most common symptoms include cognitive dysfunction (memory impairment, difficulty concentrating, attention problems, ataxia, and slow movements), psychiatric disorders (mood disorders, anxiety disorders, and psychosis), chronic fatigue, autonomic neuropathy, peripheral neuritis, and extrapyramidal symptoms.

Notable incidents involving organophosphorus nerve agents include the Tokyo Subway Sarin attack and the Matsumoto Sarin Incident, which represent the largest terrorist events using these agents on civilians. In addition to the well-documented immediate effects, long-term neuropsychiatric problems have been reported in survivors [43–45]. Gulf War Illness is a chronic disease that is believed to have affected military personnel deployed in the 1990–1991 Gulf War. A variety of neuropsychiatric symptoms have been reported, including chronic fatigue, myalgia, cognitive dysfunction, insomnia, and depression. Exposure to organophosphorus and pyridostigmine bromide has been postulated as a potential cause of toxicity [46].

Some reports have suggested that exposure to organophosphorus increases the risk of neurodegenerative diseases. Epidemiological studies have shown that occupational exposure to organophosphorus pesticides leads to long-term cognitive decline [47]. Exposure to organophosphorus has also been reported to increase the risk of developing dementia by 1.53-fold, even after adjusting for factors such as age, sex, and APOE status [48]. Furthermore, elevated serum organophosphorus levels have been associated with an increased risk of dementia, indicating that individuals with the APOE4ε4 allele may be more susceptible to organophosphorus [49]. Animal model studies [50] also suggest that exposure to organophosphorus may also cause tau accumulation. While accumulating evidence suggests a link between chronic or massive exposure to organophosphorus and neurodegenerative diseases, potentially involving tau accumulation in the brain, definitive confirmation is still pending. Organophosphorus exposure has been shown to cause persistent neuroinflammation [51], which may accelerate tau accumulation in the brain.

2   Participants

2.1   Subjects

Tau positron emission tomography (PET) studies of CTE require the recruitment of subjects with a history of repetitive mild TBI. To exclude the influence of factors other than repetitive mild TBI, such as other neuropsychiatric diseases, participants with identical clinical characteristics must be recruited. Detailed medical history and comorbidities should be obtained, and patients should be rigorously screened using operative diagnostic criteria. Age- and sex-matched healthy controls are needed to detect increased tau accumulation due to repetitive minor head trauma. This study was approved by the Institutional Review Board of the National Institutes for Quantum Science and Technology (QST), and were registered with the University Hospital Medical Information Network Clinical Trials Registry (UMIN000013517 and UMIN000030248). Written informed consent was obtained from each subject after providing a complete explanation of the study.

2.2 Clinical Diagnostic Criteria for Traumatic Encephalopathy Syndrome (TES)

CTE is primarily a neuropathological diagnosis, with definitive confirmation only possible through postmortem brain evaluation. However, the need for therapeutic intervention in high-risk individuals necessitates the development of clinical diagnostic criteria. To address this need, the concept of traumatic encephalopathy syndrome (TES) was introduced in 2014 as a clinical counterpart to neuropathologically diagnosed CTE [52]. In 2021, the National Institute of Neurological Disorders and Stroke (NINDS) revised the diagnostic criteria for TES [53]. The latest NINDS criteria for TES follow a structured approach:

1. Assessment of exposure: The first step involves determining the presence or absence of exposure to repetitive mild traumatic brain injury (TBI).

2. Identification of core clinical manifestations: The criteria require the presence of at least one of the following core features: (A) Cognitive dysfunction, (B) neurobehavioral dysregulation, and (C) progressive nature of symptoms.

3. Diagnosis: TES is diagnosed based on the presence of a "core" clinical manifestation in conjunction with a history of repetitive mild TBI.

4. Severity assessment: Once diagnosed, the severity of TES is determined.

5. Evaluation of supportive clinical features: The presence or absence of supportive clinical features is assessed to estimate the likelihood of underlying CTE pathology.

3   Methodology

3.1 Neuroimaging Biomarkers of CTE

Currently, there is no established treatment or prevention for CTE. The severity of CTE symptoms tends to worsen with increased exposure to head impacts [54]; therefore, the prevention and early detection of tau protein accumulation are the highest priorities for individuals at high risk for CTE. However, diagnosis of is currently based on autopsy, and there is no established biomarker for detecting tau accumulations in a living brain [1]. To address these limitations, there is an urgent need for neuroimaging biomarkers capable of detecting TBI-induced tau protein accumulation in living individuals. The development of reliable neuroimaging bio-markers would be invaluable for guiding early interventions, such as cognitive rehabilitation and pharmacological treatments, in individuals with CTE [55].

PET is a promising method for detecting diverse pathological changes in neurodegenerative diseases. Unlike AD, amyloid PET is not a biomarker that confirms the diagnosis of CTE, although it may be useful in differentiating CTE from AD. Neuroinflammation PET, utilizing ligands for the 18-kDa translocator protein (TSPO) to assess activated microglia and for monoamine oxidase B (MAO-B) to evaluate reactive astrocytes, has shown utility in pathophysiological studies of the long-term effects of TBI. However, this approach lacks specificity, as neuroinflammation is a common feature in various neurodegenerative diseases beyond CTE. The most important pathological change in CTE is the accumulation of p-tau; therefore, tau PET imaging holds the most promise as a biomarker of CTE.

Given that the hallmark pathological change in CTE is the accumulation of phosphorylated tau (p-tau), tau PET imaging holds the greatest promise as a potential biomarker for CTE. This imaging modality directly targets the primary neuropathological feature of the disease, potentially offering a more specific and sensitive approach to in vivo CTE diagnosis.

3.2 Development of Tau PET Tracers

Tau imaging is a molecular imaging technique that utilizes Positron Emission Tomography (PET) with specific probes to noninvasively detect pathological changes associated with tau aggregates in neurodegenerative diseases. The development of tau PET tracers has progressed through two generations, each with its own advantages and limitations.

First-Generation Tau PET Tracers:

In the 2010s, three main classes of first-generation tau PET tracers were developed:

1. PBB compounds: The ^{11}C-PBB3 tracer was shown to image tau aggregates in Alzheimer's disease (AD) and frontotemporal lobar degeneration (FTLD) [56].

2. Quinoline derivatives: A group of Tohoku University has reported the preclinical evaluation of a series of compounds, including quinoline and benzimidazole derivatives [57]. Subsequently, ^{18}F-THK5351 and other quinoline derivatives have been used for PET imaging of tau aggregates in AD [58].

3. Benzimidazole compounds: In the United States, the benzimidazole compound BF-128 was structurally modified, leading to the identification of novel tau PET probes called T-807 and T-808 [59, 60], which have been shown to detect tau deposits in AD [59, 61].

Although these three strains of tau PET probes are collectively referred to as first-generation tau PET tracers, potential problems exist regarding their selectivity for tau aggregates and metabolic properties. ^{18}F-THK5351 and analogous quinoline compounds show significant off-target binding to monoamine oxidase-B (MAO-B) [62], and ^{18}F-flortaucipir shows cross-reactivity with neuronal melanin-containing neurons, meningeal melanocytes, monoamine oxidase-A (MAO-A). In a previous study, ^{11}C-PBB3 did not bind to either MAO-A or MAO-B (10) but was rapidly converted to a metabolite [63], resulting in insufficient brain uptake [56]. In addition, ^{18}F-flortaucipir and ^{11}C-PBB3 show nonspecific accumulation of radioactivity in the choroid plexus [56, 64], and high radioactivity retention is observed in the venous sinus, probably because of the radioactive metabolite of ^{11}C-PBB3 [56, 64]. These radioactive signals, which do not reflect tau deposition, contribute to reduced image contrast in PET visualization of tau deposition.

To address the limitations of first-generation tracers, second-generation tau PET tracers have been developed with the aims of reducing off-target binding, minimizing nonspecific accumulation, and increasing the signal-to-background ratio. Currently, most of the clinically available second-generation tau PET tracers are derivatives of ^{18}F-flortaucipir, such as ^{18}F-MK-6240 [65], ^{18}F-RO-948 [66], ^{18}F-PI-2620 [67], and ^{18}F-GTP1 [68]. Another quinoline derivative, ^{18}F-JNJ-067, has also been evaluated in humans; however, a substantial off-target signal, presumably from MAO-B, was observed [69]. In contrast, ^{18}F-florzolotau, also known as ^{18}F-APN1607/^{18}F-PM-PBB3, was proven to have favorable properties, such as metabolic stability, high brain uptake, and no significant radioactive signal accumulation in the venous sinus, and was found to be highly sensitive in detecting AD and non-AD tau aggregates [70]. The development of these second-generation tracers represents a significant advancement in tau PET imaging, potentially offering improved specificity and sensitivity for detecting tau pathology in various neurodegenerative disorders, including CTE.

3.3 Tau PET Imaging Study of CTE with the First-Generation Tau PET Tracer

We performed a tau PET imaging study using ^{11}C-PBB3 in subjects with a history of mild-repetitive and severe TBI [71]. PET data were acquired from 27 subjects (mean $\pm$ SD: 44.8 $\pm$ 12.5 years) in the long-term survivors of mild-repetitive ($n = 13$) or severe TBI ($n = 14$) and from 15 healthy control subjects (mean $\pm$ SD: 43.4 $\pm$ 14.4 years). The mild-repetitive TBI group had prolonged exposure to concussions from contact sports (mean $\pm$ SD: 16.5 $\pm$ 9.6 years; min–max: 3–29 years), with an average of 28.3 $\pm$ 11.7 years since the first head injury. The severe TBI group consisted of patients who were injured in traffic accidents ($n = 12$) or falls ($n = 2$). The average period after head injury was

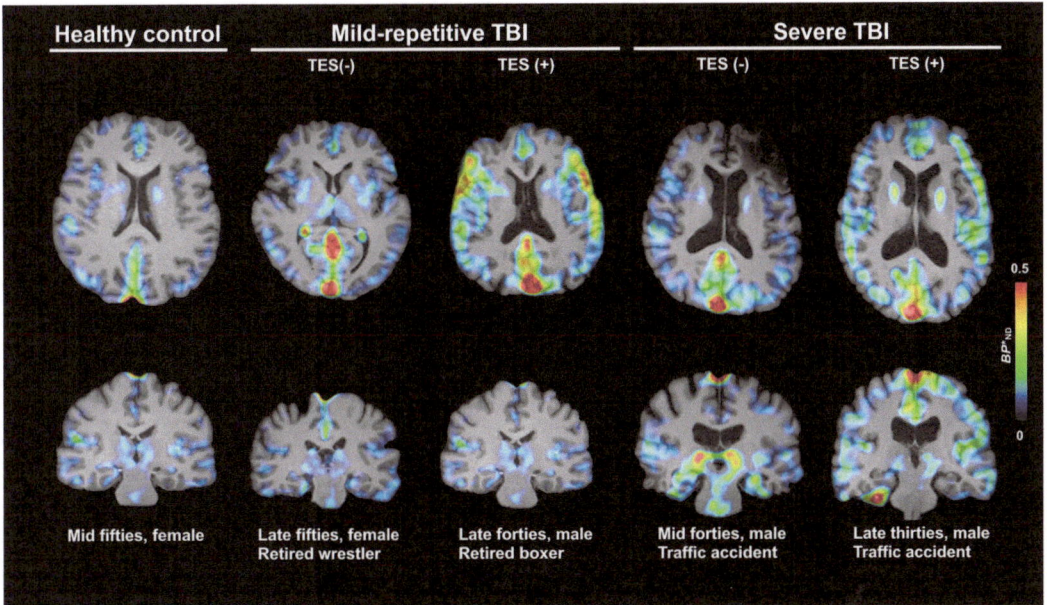

Fig. 3 Representative positron emission tomography (PET) images of first-generation tau tracer in long-term survivors of mild-repetitive and severe traumatic brain injury (TBI). *Binding potential* $(BP)^*_{ND}$ parametric images are shown in five representative cases. The color bar shows BP^*_{ND} values ranging from 0 to 0.5. Each PET image was overlaid onto individual magnetic resonance (MR) T1 images. Patients with TBI and traumatic encephalopathy syndrome (TES) exhibited increased BP^*_{ND} values in several widespread brain regions

14.5 ± 12.1 years, and the average duration of loss of consciousness (LOC) following the injury was 14.1 ± 12.2 days.

Late-onset neuropsychiatric symptoms following TBI were diagnosed based on the modified criteria for traumatic encephalopathy syndrome (TES) [52]. Representative [1]C-PBB3-PET images are shown in Fig. 2. Parametric images of ^{11}C-PBB3 binding potential (BP^*_{ND}) showed minimum difference in the tracer retention pattern between healthy control subjects and TBI patients without TES. However, both mild-repetitive and severe TBI patients with TES exhibited significantly higher BP^*_{ND} values, primarily in the frontal, parietal, and temporal lobes (Fig. 3). Furthermore, a relationship was found between the severity of delayed symptoms and TBI, and the total number of tau lesions with 11C-PBB3, indicating the utility of tau PET with ^{11}C-PBB3 in TBI.

To validate ^{11}C-PBB3 binding to tau deposits in TBI brains, we performed in vitro phospho-tau antibody and PBB3 fluorescence labeling of CTE brain sections. The in vitro immunofluorescence assays revealed PBB3-positive tau inclusions at the depths of neocortical sulci, confirming ^{11}C-PBB3 binding to tau lesions (Fig. 4).

While our findings demonstrate that CTE-type tau aggregates can be detected using first-generation tau PET tracers, several technical limitations persist. These include nonspecific

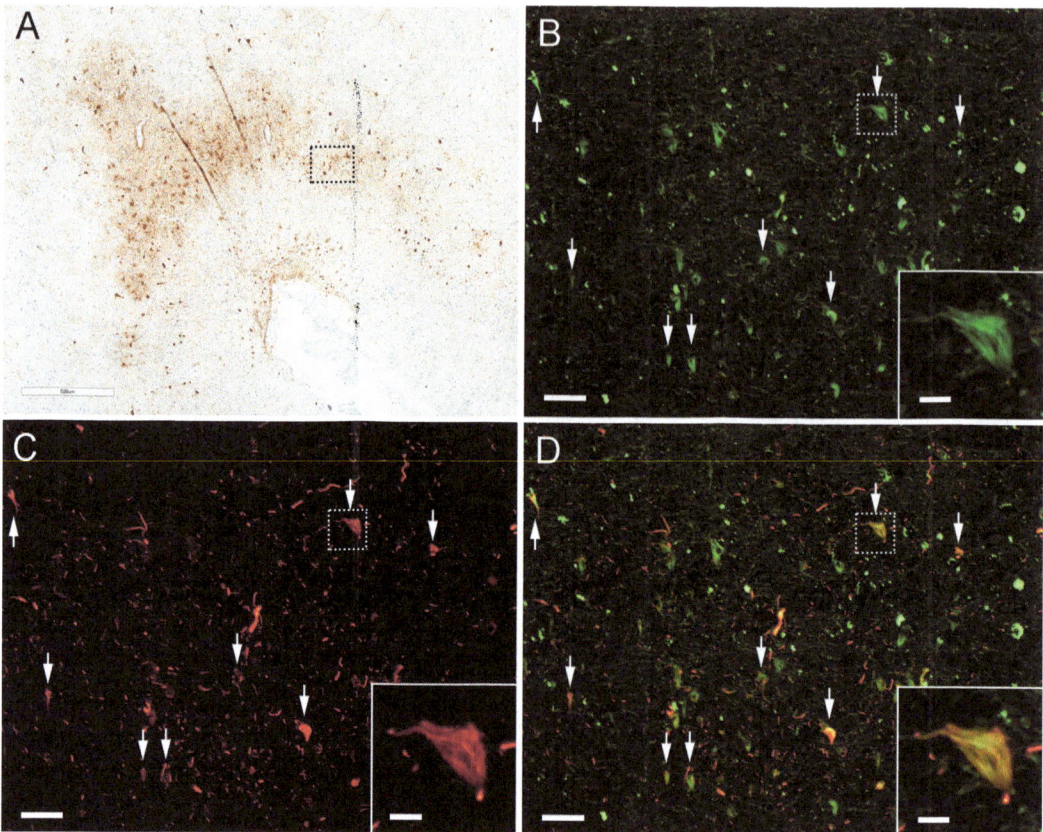

Fig. 4 Tau immunohistochemistry and PBB3 fluorescence labeling in postmortem brain tissue of patients with chronic traumatic encephalopathy (CTE). Immunohistochemical staining with phospho-tau antibodies CP13 (A), PHF1 (C), and histochemical labeling with PBB3 (**b**) were performed on sections of the superior temporal gyrus in case 1. Panel A depicts CP13 antibody-labeled abundant neurofibrillary tangles (NFTs) at the depth of the sulcus. TSAs were also found in the sulcus of the superior temporal gyrus and perivascular regions. In panels B–D, the same sections (black-dotted box in **a**) were stained with PBB3 and PHF1. Panel B depicts PBB3 fluorescence labeling that revealed the presence of several NFTs. Panel C depicts thePHF1-positive structures including NFTs and threads that were observed in the same region as in Panel B. Panel D depicts the merged images showing labeling of NFTs with both PBB3 and PHF1 antibodies. Arrows show the sites of neuronal inclusions stained with both PBB3 and PHF1. The insets in panels B–D show a tangled structure that was positive for both PBB3 and PHF1. The white-dotted boxes in panels B–D indicate the structure estimated as a ghost tangle. Bar in panel A = 500 μm. Bars in panels B–D = 50 μm. Bars in insets of panels B–D = 20 μm. PHF1: paired helical filament-1

accumulation in the venous sinus, off-target binding to the choroid plexus and basal ganglia, low image contrast due to a suboptimal signal-to-noise ratio, and metabolic instability. Moreover, the unique distribution pattern of tau lesions in CTE necessitates multiple neuroimaging modifications to accurately quantify tau lesions using ^{11}C-PBB3.

3.4 Quantification of CTE Tau Burden with the First-Generation Tau PET Tracer

The standard uptake value ratio (SUVR) method, commonly used to quantify tau lesions in Alzheimer's disease (AD) and other neurodegenerative diseases, is inadequate for assessing tau pathology in TBI. This method, which uses cerebellar gray matter as a reference region, presents two main challenges in TBI contexts:

1. Potential underestimation due to cerebellar tau accumulation in CTE, unlike in AD where cerebellar tau is minimal.

2. Inaccurate capture of TBI-induced tau accumulation due to the small, sparsely distributed (patchy) nature of tau lesions, which can be obscured by partial volume effects when comparing SUVR within regions of interest (ROIs).

To address these problems, we developed a new quantitative method.

Optimized reference voxel selection and BP^*_{ND} image generation:

1. Initial BP^*_{ND} parametric images were generated using an original multilinear reference tissue model (MRTMO) [72], employing the cerebellar cortex as the reference region.

2. Optimized reference voxels were then extracted [73, 74], based on the mean full width at half maximum (FWHM) of a control histogram from age-matched healthy controls.

3. Using this optimized reference tissue, we reconstructed parametric BP^*_{ND} images using $MRTM_O$.

Quantification of Sparsely Distributed Tau Lesions:

We developed an index that conjunctively reflects the intensity and volume of PET signals, adapting techniques used in other PET studies This technique has been applied to the calculation of total cerebral glucose metabolism and accurate tumor size in PET studies with ^{18}F-fluoro-deoxy-glucose (^{18}F-FDG) [75] and the quantification of pancreatic islet cell signals in ^{11}C-dihydrotetrabenazine-PET [76]. This technique calculates the mean of PET measures (e.g., BP_{ND}, SUV, and SUVR) multiplied by segmented brain volumes as binding capacity and has also been employed for the analysis of ^{18}F-FDG data in the brains of patients with AD [77]. Our ^{11}C-PBB3 binding capacity was measured as follows:

1. Thresholds for BP^*_{ND} were defined separately for gray matter (GM) and white matter (WM) segments as one standard deviation from the mean value in healthy subjects.

2. ^{11}C-PBB3 binding capacity in each segment of interest was quantified as the sum of the individual voxels BP^*_{ND} above the threshold, multiplied by the voxel volume:

$$^{11}\text{C-PBB3 binding capacity}$$

$$= \sum \left(\text{individual voxel } BP^*_{\text{ND}} \times \text{Voxel volume}\right)$$

The unit is cm^3 (volume) because BP^*_{ND} is unitless.

Our analysis revealed that patients with TBI showed increased $^{11}\text{C-PBB3}$ binding capacity in widespread brain regions compared to age-matched HC subjects [71]. Patients with TBI and traumatic encephalopathy syndrome (TES) exhibited higher $^{11}\text{C-PBB3}$ binding capacity in the WM segment, particularly in the frontal WM, compared to those without TES. This increased binding capacity was primarily due to elevated radio signals in the surface white matter. Notably, the distribution pattern of the tau tracer in our study was unique to TBI and distinct from that observed in AD. We found a positive correlation between $^{11}\text{C-PBB3}$ binding capacity in the WM segment and the severity of psychosis, as assessed by the Brief Psychiatric Rating Scale (BPRS). These findings suggest a strong association between tau neuropathology and late-onset neuropsychiatric symptoms following TBI.

In conclusion, our novel quantification method addresses the specific challenges of assessing tau pathology in TBI with the first-generation tau PET tracer and provides valuable insights into the relationship between tau accumulation and clinical symptoms in CTE.

3.5 Tau PET Imaging of CTE with the Second-Generation Tau PET Tracer

The second-generation tau PET tracer $^{18}\text{F-florzolotau}$ can visualize tau lesions caused by TBI with higher contrast than first-generation tau PET drugs. An overview of our study is shown in Fig. 5. Our longitudinal cohort study included individuals with a history of long-term exposure to mild-repetitive TBI, such as former professional boxers. Each subject underwent clinical evaluation, neuroimaging assessment by tau PET, and blood biomarker testing approximately every 2 years.

The methods used to analyze the PET images are shown in Fig. 6. Considering the possibility that there are no brain regions in CTE brains that are intact due to tau accumulation, we established a new method to extract optimized reference regions from the gray matter [78]. In the present study, the regions of interest were the gray matter, gray-white matter boundary, white matter, and subcortical regions.

Representative images of $^{18}\text{F-florzolotau}$ in two former athletes with a prolonged exposure to concussions are shown in Fig. 7. Case 1 is a male patient in his early 60's, who spent nearly 40 years playing ice hockey. He began his ice hockey career at age 15 and retired at age 55. He had experienced multiple concussions, including severe incidents at ages 50 and 55. At the age of 57, he was also involved in a car accident, with his head striking the front windshield. The first indication of psychiatric problems occurred in his

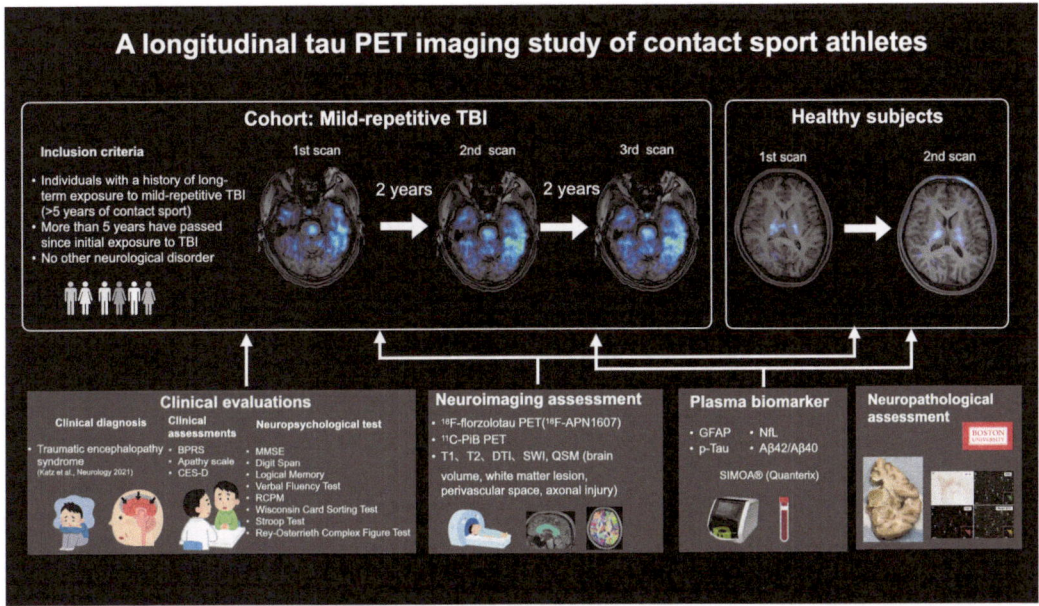

Fig. 5 A longitudinal tau positron emission tomography (PET) imaging study of contact sport athletes. Our longitudinal cohort study included individuals with a history of long-term exposure to mild-repetitive traumatic brain injury (TBI). Each subject underwent clinical evaluation, neuroimaging assessment, and blood biomarker testing approximately every 2 years

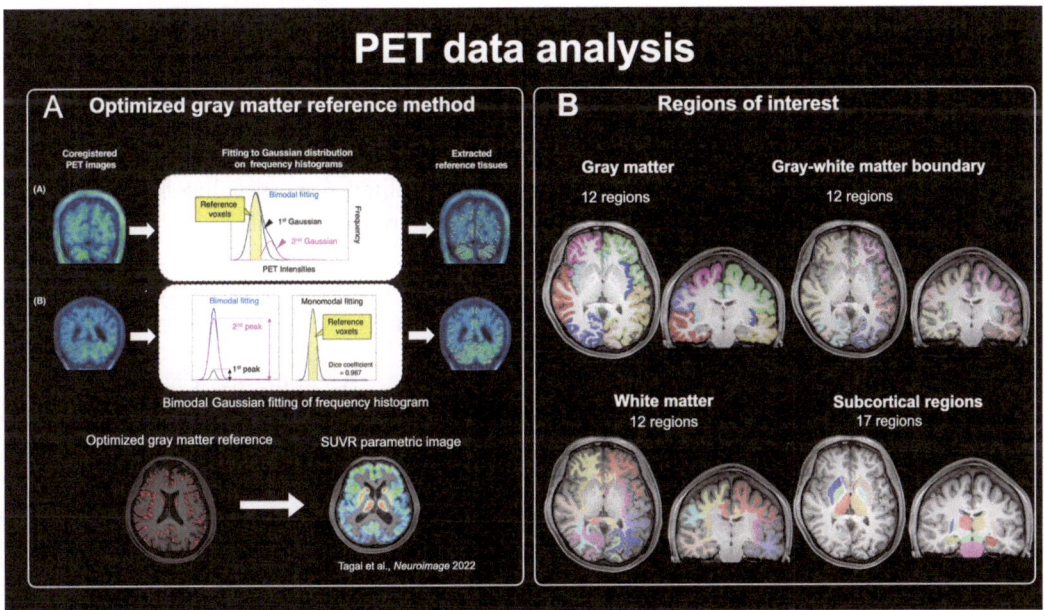

Fig. 6 PET data analysis. Considering the possibility that there are no intact brain regions in chronic traumatic encephalopathy (CTE) because of tau accumulation, we established a new method to extract optimized reference regions from the gray matter [78]. The regions of interest were the gray matter, gray-white matter boundary, white matter, and subcortical regions

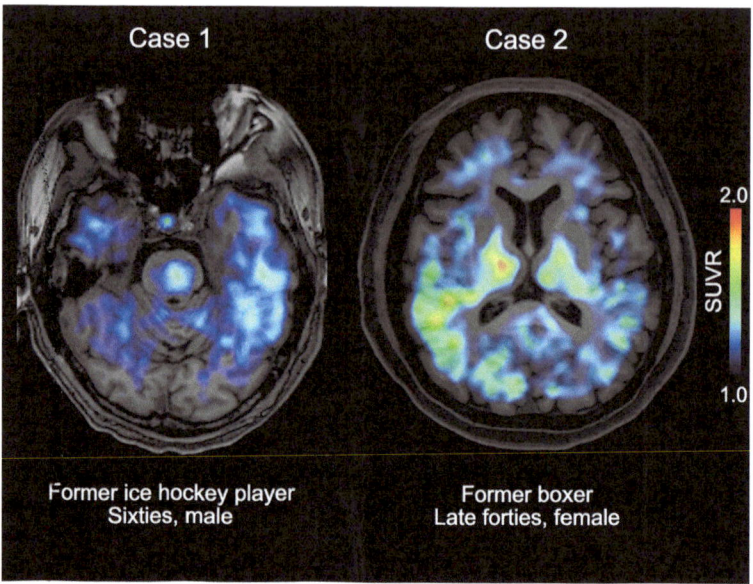

Fig. 7 Representative positron emission tomography (PET) images of second-generation tau tracer in long-term survivors of mild-repetitive and severe traumatic brain injury (TBI). Case 1: focal accumulation of ^{18}F-florzolotau was observed in the right inferior temporal lobe. Case 2: focal accumulation of ^{18}F-florzolotau was observed in the left occipital and parietal lobes

mid-50s. At the age of 56 years, he complained of insomnia, depression, irritability, and hypersensitivity. He also experienced tinnitus, severe headaches, and dizziness. Depression persisted for years, and suicidal ideation sometimes occurred. At the age of 61 years, the patient showed a manic episode. He suddenly stood as a candidate for the metropolitan governorship but was unsuccessful. The amyloid PET was negative, indicating the absence of Aβ pathology. ^{18}F-florzolotau PET showed focal tracer accumulation in the left inferior temporal lobe. The tracer distribution patterns differed from those in AD and other neurodegenerative diseases.

Case 2 is a woman in her late 40 s, who was a former domestic champion of female boxing. She began boxing at the age of 23 years. Since there was no official federation for female professional boxers in Japan at the time, she had been sparring with male professional boxers for training. At the age of 30 years, she received a strong blow to the right temple and consequently hit her head on a concrete wall. Within a few years after her first fight, she won a domestic amateur championship. In her last fight, she was knocked out by her opponent and was transported to a hospital by ambulance. At approximately 30 years of age, she developed depression and experienced severe weight loss. She also demonstrated psychotic symptoms such as auditory hallucinations and delusional thoughts, she was diagnosed with schizophrenia. At the age of 35, she retired from boxing because she felt ataxic and had difficulty walking. She was hospitalized several times because of the recurrence of psychotic symptoms. The amyloid PET was negative. ^{18}F-florzolotau PET showed focal accumulation of ^{18}F-florzolotau in the right parietal lobe, and occipital lobe and bilateral basal ganglia.

Although this needs to be verified through correlation with imaging pathology, both of these cases show atypical non-AD-type patterns of tau accumulation, and it is presumed that they exhibit pathology corresponding to CTE. The PET findings in these two cases indicate that tau PET scans are useful for the diagnosis of neurodegenerative diseases caused by TBI.

3.6 Tau PET Imaging of Long-Term Survivor of Organophosphate Poisoning

Although it has been suggested that organophosphate poisoning may cause long-term accumulation of tau in the brain, there was no clinical evidence to support this. Here, we present the tau PET images of a patient with a history of organophosphorus poisoning. The patient was diagnosed with bipolar disorder at the age of 60 and was thereafter treated with mood stabilizers. At the age of 63 years, his depressive symptoms worsened, and he attempted suicide by oral administration of a pesticide containing organophosphorus. After 3 weeks of emergency care in the intensive care unit, his general condition recovered. Cognitive function was mostly preserved; however, memory impairment was prominent. Subsequently, the cognitive dysfunction gradually progressed. At the age of 75, 12 years after the pesticide poisoning, PET with ^{18}F-florzolotau was performed. The tau PET showed focal ^{18}F-florzolotau uptake in the occipital and temporal cortices, and in subcortical regions with left-sided dominance, as shown in Fig. 8. The distribution of ^{18}F-florzolotau radio signals was distinct from that of AD, PSP, or other FTLDs, and the amyloid PET results were negative. In this case, the causal relationship between organophosphate poisoning and tau accumulation in the brain is not clear. Accumulating more case data will likely reveal the exact relationship between organophosphate exposure and tauopathy.

3.7 Future Directions of Research

[18]F-florzolotau, a second-generation tau PET tracer, has the potential for the early diagnosis of CTE. It is also expected to allow diagnosis at the single-case level owing to its improved image contrast, metabolic stability, and reduced off-target binding compared with first-generation tau PET tracers. The high degree of accuracy in the visualization of tau lesions in CTE brains may also be useful for developing new therapies targeting tau lesions caused by TBI. Verification using image-pathological correlation analysis in the same individual is essential to confirm the accuracy of PET in detecting tau aggregates in CTE.

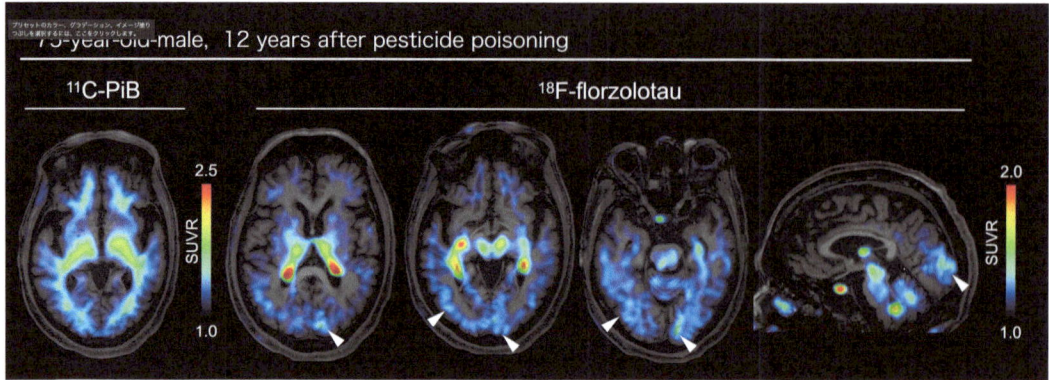

Fig. 8 Representative positron emission tomography (PET) of tauopathy following organophosphorus pesticide intoxication. This figure depicts a neuroimaging finding of the case of long-term survivor of organophosphorus pesticide poisoning. The patient was admitted to the intensive care unit (ICU) for severe organophosphorus poisoning. PET with ^{11}C-PiB and ^{18}F-florzolotau was performed 12 years after exposure. Tau PET showed focal ^{18}F-florzolotau uptakes in the occipital and temporal cortices and subcortical regions, with left-side dominance, as shown in the figure. The distribution of ^{18}F-florzolotau radio signals was distinct from that of Alzheimer's disease, progressive supranuclear palsy, or other frontotemporal lobe degeneration (FTLDs), and amyloid PET results were negative

References

1. McKee AC, Stern RA, Nowinski CJ et al (2013) The spectrum of disease in chronic traumatic encephalopathy. Brain 136:43–64

2. Omalu BI, DeKosky ST, Minster RL et al (2005) Chronic traumatic encephalopathy in a National Football League player. Neurosurgery 57:128–134; discussion 128–34

3. Langworth-Green C, Patel S, Jaunmuktane Z et al (2023) Chronic effects of inflammation on tauopathies. Lancet Neurol 22:430–442

4. Mez J, Daneshvar DH, Kiernan PT et al (2017) Clinicopathological evaluation of chronic traumatic encephalopathy in players of American football. JAMA 318:360–370

5. Stewart W, Buckland ME, Abdolmohammadi B et al (2023) Risk of chronic traumatic encephalopathy in rugby union is associated with length of playing career. Acta Neuropathol 146:829–832

6. van Amerongen S, Kamps S, Kaijser KKM et al (2023) Severe CTE and TDP-43 pathology in a former professional soccer player with dementia: a clinicopathological case report and review of the literature. Acta Neuropathol Commun 11:77

7. Omalu BI, Fitzsimmons RP, Hammers J, Bailes J (2010) Chronic traumatic encephalopathy in a professional American wrestler. J Forensic Nurs 6:130–136

8. Schwab N, Wennberg R, Grenier K et al (2021) Association of position played and career

duration and chronic traumatic encephalopathy at autopsy in elite football and hockey players. Neurology 96:e1835–e1843

9. McKee AC, Daneshvar DH, Alvarez VE, Stein TD (2014) The neuropathology of sport. Acta Neuropathol 127:29–51

10. Takahata K, Kubota M, Tagai K et al (2023) The aftermath of boxing: longitudinal changes of cerebral retention patterns of ^{18}F-florzolotau (^{18}F-APN-1607) in the former professional boxers. Alzheimers Dement 19. https://doi.org/10.1002/alz.076455

11. Priemer DS, Iacono D, Rhodes CH et al (2022) Chronic traumatic encephalopathy in the brains of military personnel. N Engl J Med 386:2169–2177

12. Tai XY, Koepp M, Duncan JS et al (2016) Hyperphosphorylated tau in patients with refractory epilepsy correlates with cognitive decline: a study of temporal lobe resections. Brain 139:2441–2455

13. Lee K, Kim S-I, Lee Y et al (2017) An autopsy proven child onset chronic traumatic encephalopathy. Exp Neurobiol 26:172–177

14. Danielsen T, Hauch C, Kelly L, White CL (2021) Chronic traumatic encephalopathy (CTE)-type neuropathology in a young victim of domestic abuse. J Neuropathol Exp Neurol 80:624–627

15. McKee AC, Mez J, Abdolmohammadi B et al (2023) Neuropathologic and clinical findings

in young contact sport athletes exposed to repetitive head impacts. JAMA Neurol. https://doi.org/10.1001/jamaneurol.2023.2907

16. Stein TD, Alvarez VE, McKee AC (2014) Chronic traumatic encephalopathy: a spectrum of neuropathological changes following repetitive brain trauma in athletes and military personnel. Alzheimers Res Ther 6:4

17. Stern RA, Daneshvar DH, Baugh CM et al (2013) Clinical presentation of chronic traumatic encephalopathy. Neurology 81:1122–1129

18. Iverson GL (2016) Suicide and chronic traumatic encephalopathy. J Neuropsychiatry Clin Neurosci 28:9–16

19. Mahar I, Alosco ML, McKee AC (2017) Psychiatric phenotypes in chronic traumatic encephalopathy. Neurosci Biobehav Rev 83:622–630

20. Nicks R, Clement NF, Alvarez VE et al (2023) Repetitive head impacts and chronic traumatic encephalopathy are associated with TDP-43 inclusions and hippocampal sclerosis. Acta Neuropathol 145:395–408

21. Takahata K, Tabuchi H, Mimura M (2016) Late-onset neurodegenerative diseases following traumatic brain injury: chronic traumatic encephalopathy (CTE) and Alzheimer's disease secondary to TBI (AD-TBI). Brain Nerve 68:849–857

22. Johnson VE, Stewart W (2015) Traumatic brain injury: age at injury influences dementia risk after TBI. Nat Rev Neurol 11:128–130

23. Fujii DE, Ahmed I (2001) Risk factors in psychosis secondary to traumatic brain injury. J Neuropsychiatry Clin Neurosci 13:61–69

24. Fujii D, Ahmed I (2002) Characteristics of psychotic disorder due to traumatic brain injury: an analysis of case studies in the literature. J Neuropsychiatry Clin Neurosci 14:130–140

25. Fujii D, Fujii DC (2012) Psychotic disorder due to traumatic brain injury: analysis of case studies in the literature. J Neuropsychiatry Clin Neurosci 24:278–289

26. Sachdev P, Smith JS, Cathcart S (2001) Schizophrenia-like psychosis following traumatic brain injury: a chart-based descriptive and case-control study. Psychol Med 31:231–239

27. Takahata K, Kato M, Mimura M, Shimada H (2016) Late-onset neurocognitive deficits following traumatic brain injury: chronic traumatic encephalopathy (CTE) and psychotic disorder following TBI (PDF TBI). Higher Brain Funct Res 35:276–282

28. Roberts GW, Allsop D, Bruton C (1990) The occult aftermath of boxing. J Neurol Neurosurg Psychiatry 53:373–378

29. Ikonomovic MD, Uryu K, Abrahamson EE et al (2004) Alzheimer's pathology in human temporal cortex surgically excised after severe brain injury. Exp Neurol 190:192–203

30. Roberts GW, Gentleman SM, Lynch A et al (1994) Beta amyloid protein deposition in the brain after severe head injury: implications for the pathogenesis of Alzheimer's disease. J Neurol Neurosurg Psychiatry 57:419–425

31. Johnson VE, Stewart W, Smith DH (2012) Widespread τ and amyloid-β pathology many years after a single traumatic brain injury in humans. Brain Pathol 22:142–149

32. Okamura Y, Kawakami I, Watanabe K et al (2019) Tau progression in single severe frontal traumatic brain injury in human brains. J Neurol Sci 407:116495

33. Goldstein LE, Fisher AM, Tagge CA et al (2012) Chronic traumatic encephalopathy in blast-exposed military veterans and a blast neurotrauma mouse model. Sci Transl Med 4:134ra60

34. Smith DH, Johnson VE, Stewart W (2013) Chronic neuropathologies of single and repetitive TBI: substrates of dementia? Nat Rev Neurol 9:211–221

35. Ikeda K, Akiyama H, Kondo H et al (1995) Numerous glial fibrillary tangles in oligodendroglia in cases of subacute sclerosing panencephalitis with neurofibrillary tangles. Neurosci Lett 194:133–135

36. Qi C, Hasegawa M, Takao M et al (2023) Identical tau filaments in subacute sclerosing panencephalitis and chronic traumatic encephalopathy. Acta Neuropathol Commun 11:74

37. Qi C, Verheijen BM, Kokubo Y et al (2023) Tau filaments from amyotrophic lateral sclerosis/parkinsonism-dementia complex adopt the CTE fold. Proc Natl Acad Sci USA 120:e2306767120

38. Niklowitz WJ, Mandybur TI (1975) Neurofibrillary changes following childhood lead encephalopathy. J Neuropathol Exp Neurol 34:445–455

39. Pollanen MS, Onzivua S, McKeever PM et al (2023) The spectrum of disease and tau pathology of nodding syndrome in Uganda. Brain 146:954–967

40. Theis H, Bischof GN, Brüggemann N et al (2023) In vivo measurement of tau depositions in anti-IgLON5 disease using 18F-PI-2620 PET. Neurology. https://doi.org/10.1212/WNL.0000000000207870

41. Gelpi E, Höftberger R, Graus F et al (2016) Neuropathological criteria of anti-IgLON5-

related tauopathy. Acta Neuropathol 132:531–543

42. Jokanović M (2018) Neurotoxic effects of organophosphorus pesticides and possible association with neurodegenerative diseases in man: a review. Toxicology 410:125–131

43. Hatta K, Miura Y, Asukai N, Hamabe Y (1996) Amnesia from sarin poisoning. Lancet 347: 1343

44. Yanagisawa N, Morita H, Nakajima T (2006) Sarin experiences in Japan: acute toxicity and long-term effects. J Neurol Sci 249:76–85

45. Hoffman A, Eisenkraft A, Finkelstein A et al (2007) A decade after the Tokyo sarin attack: a review of neurological follow-up of the victims. Mil Med 172:607–610

46. Chao LL, Kriger S, Buckley S et al (2014) Effects of low-level sarin and cyclosarin exposure on hippocampal subfields in Gulf War Veterans. Neurotoxicology 44:263–269

47. Bosma H, van Boxtel MPJ, Ponds R et al (2000) Pesticide exposure and risk of mild cognitive dysfunction. Lancet 356:912–913

48. Hayden KM, Norton MC, Darcey D et al (2010) Occupational exposure to pesticides increases the risk of incident AD: the Cache County study. Neurology 74:1524–1530

49. Richardson JR, Roy A, Shalat SL et al (2014) Elevated serum pesticide levels and risk for Alzheimer disease. JAMA Neurol 71:284–290

50. Yates PL, Case K, Sun X et al (2022) Veteran-derived cerebral organoids display multifaceted pathological defects in studies on Gulf War Illness. Front Cell Neurosci 16. https://doi.org/10.3389/fncel.2022.979652

51. Flannery BM, Bruun DA, Rowland DJ et al (2016) Persistent neuroinflammation and cognitive impairment in a rat model of acute diisopropylfluorophosphate intoxication. J Neuroinflammation 13:1–16

52. Montenigro PH, Baugh CM, Daneshvar DH et al (2014) Clinical subtypes of chronic traumatic encephalopathy: literature review and proposed research diagnostic criteria for traumatic encephalopathy syndrome. Alzheimers Res Ther 6:68

53. Katz DI, Bernick C, Dodick DW et al (2021) National Institute of Neurological Disorders and Stroke consensus diagnostic criteria for traumatic encephalopathy syndrome. Neurology 96:848–863

54. Daneshvar DH, Nair ES, Baucom ZH et al (2023) Leveraging football accelerometer data to quantify associations between repetitive head impacts and chronic traumatic encephalopathy in males. Nat Commun 14:3470

55. Takahata K, Higuchi M, Mimura M (2022) Long-term sequelae of mild-repetitive and severe traumatic brain injury: clinical manifestations, neuropathology and diagnosis by tau PET imaging. In: Rajendram R (ed) Neuroscience of traumatic brain injury. Academic Press

56. Maruyama M, Shimada H, Suhara T et al (2013) Imaging of tau pathology in a tauopathy mouse model and in Alzheimer patients compared to normal controls. Neuron 79: 1094–1108

57. Okamura N, Suemoto T, Furumoto S et al (2005) Quinoline and benzimidazole derivatives: candidate probes for in vivo imaging of tau pathology in Alzheimer's disease. J Neurosci 25:10857–10862

58. Harada R, Okamura N, Furumoto S et al (2016) ^{18}F-THK5351: a novel PET radiotracer for imaging neurofibrillary pathology in Alzheimer disease. J Nucl Med 57:208–214

59. Chien DT, Bahri S, Szardenings AK et al (2013) Early clinical PET imaging results with the novel PHF-tau radioligand [F-18]-T807. J Alzheimers Dis 34:457–468

60. Chien DT, Szardenings AK, Bahri S et al (2014) Early clinical PET imaging results with the novel PHF-tau radioligand [F18]-T808. J Alzheimers Dis 38:171–184

61. Johnson KA, Schultz A, Betensky RA et al (2016) Tau positron emission tomographic imaging in aging and early Alzheimer disease. Ann Neurol 79:110–119

62. Ng KP, Pascoal TA, Mathotaarachchi S et al (2017) Monoamine oxidase B inhibitor, selegiline, reduces 18F-THK5351 uptake in the human brain. Alzheimers Res Ther 9:25

63. Hashimoto H, Kawamura K, Takei M et al (2015) Identification of a major radiometabolite of [^{11}C]PBB3. Nucl Med Biol 42:905–910

64. Ikonomovic MD, Abrahamson EE, Price JC et al (2016) [F-18]AV-1451 positron emission tomography retention in choroid plexus: more than "off-target" binding. Ann Neurol 80: 307–308

65. Aguero C, Dhaynaut M, Normandin MD et al (2019) Autoradiography validation of novel tau PET tracer [F-18]-MK-6240 on human postmortem brain tissue. Acta Neuropathol Commun 7:37

66. Honer M, Gobbi L, Knust H et al (2018) Preclinical evaluation of ^{18}F-RO6958948, ^{11}C-RO6931643, and ^{11}C-RO6924963 as

novel PET radiotracers for imaging tau aggregates in Alzheimer disease. J Nucl Med 59: 675–681

67. Kroth H, Oden F, Molette J et al (2019) Discovery and preclinical characterization of [18F] PI-2620, a next-generation tau PET tracer for the assessment of tau pathology in Alzheimer's disease and other tauopathies. Eur J Nucl Med Mol Imaging 46:2178–2189

68. Sanabria Bohórquez S, Marik J, Ogasawara A et al (2019) [^{18}F]GTP1 (Genentech tau probe 1), a radioligand for detecting neurofibrillary tangle tau pathology in Alzheimer's disease. Eur J Nucl Med Mol Imaging 46:2077–2089

69. Baker SL, Provost K, Thomas W et al (2021) Evaluation of [^{18}F]-JNJ-64326067-AAA tau PET tracer in humans. J Cereb Blood Flow Metab 41:3302–3313

70. Tagai K, Ono M, Kubota M et al (2021) High-contrast in vivo imaging of tau pathologies in Alzheimer's and non-Alzheimer's disease tauopathies. Neuron 109:42–58.e8

71. Takahata K, Kimura Y, Sahara N et al (2019) PET-detectable tau pathology correlates with long-term neuropsychiatric outcomes in patients with traumatic brain injury. Brain 142:3265–3279

72. Ichise M, Ballinger JR, Golan H et al (1996) Noninvasive quantification of dopamine D2 receptors with iodine-123-IBF SPECT. J Nucl Med 37:513–520

73. Kimura Y, Endo H, Ichise M et al (2016) A new method to quantify tau pathologies with (11)C-PBB3 PET using reference tissue voxels extracted from brain cortical gray matter. EJNMMI Res 6:24

74. Endo H, Shimada H, Sahara N et al (2019) In vivo binding of a tau imaging probe, [^{11}C] PBB3, in patients with progressive supranuclear palsy. Mov Disord 34:744–754

75. Vanhove K, Mesotten L, Heylen M et al (2018) Prognostic value of total lesion glycolysis and metabolic active tumor volume in non-small cell lung cancer. Cancer Treat Res Commun 15:7–12

76. Goland R, Freeby M, Parsey R et al (2009) 11C-dihydrotetrabenazine PET of the pancreas in subjects with long-standing type 1 diabetes and in healthy controls. J Nucl Med 50:382–389

77. Alavi A, Newberg AB, Souder E, Berlin JA (1993) Quantitative analysis of PET and MRI data in normal aging and Alzheimer's disease: atrophy weighted total brain metabolism and absolute whole brain metabolism as reliable discriminators. J Nucl Med 34:1681–1687

78. Tagai K, Ikoma Y, Endo H et al (2022) An optimized reference tissue method for quantification of tau protein depositions in diverse neurodegenerative disorders by PET with ^{18}F-PM-PBB3 (^{18}F-APN-1607). NeuroImage 264:119763

Chapter 5

Amyloid PET Imaging and Quantification of Amyloid Deposition

Hiroshi Matsuda and Tensho Yamao

Abstract

The diagnosis of Alzheimer's disease requires evidence of amyloid deposition in the brain. Amyloid positron emission tomography is very useful for this purpose, and its clinical importance has increased with the approval of disease-modifying drugs. The diagnosis of amyloid positivity is based on visual assessment, but small amounts of amyloid deposition are often challenging to diagnose. Quantitative assessment methods have been developed to aid in this visual diagnosis, and the Centiloid scale, a 100-point scale representing amyloid deposition regardless of the type of amyloid tracer used or the reference site required for quantification, is becoming the standard. Several software packages have also been developed to quantify amyloid deposition.

Key words Alzheimer's disease, Amyloid, Positron emission tomography, Centiloid

1 Introduction

Alzheimer's disease (AD) is an irreversible, progressive brain disorder that slowly compromises the healthy brain, affecting memory, cognitive skills, emotions, behavior, and mood. Over time, a person's ability to carry out daily activities becomes impaired. AD is the most common form of dementia, accounting for 60–80% of cases [1]. According to the 2023 World Alzheimer Report [2], the number of people living with dementia will increase from 55 million in 2019 to 139 million by 2050. In Japan, the Ministry of Health, Labour, and Welfare estimates that 7 million Japanese will have dementia by 2025, equivalent to about one in five of Japan's elderly population.

The pathophysiological process of AD is believed to begin many years before the diagnosis of AD dementia. The National Institute on Aging and the Alzheimer's Association convened an international working group to review biomarker, epidemiological, and neuropsychological evidence and develop recommendations to identify the factors that best predict the risk of progression from

Daichi Sone (ed.), *Molecular Imaging for Brain Diseases*, Neuromethods, vol. 222, https://doi.org/10.1007/978-1-0716-4494-2_5,
© The Author(s), under exclusive license to Springer Science+Business Media, LLC, part of Springer Nature 2025

normal cognition to mild cognitive impairment (MCI) and AD dementia [3, 4]. Based on the prevailing scientific evidence, as well as the testing and refinement of these models through longitudinal clinical research studies, a conceptual framework and operational research criteria were proposed [5, 6].

Patients who have memory impairment but limited functional impairment and do not meet the clinical criteria for dementia are classified as having MCI [7, 8] or mild neurocognitive disorder as defined in the *Diagnostic and Statistical Manual of Mental Disorders*, Fifth Edition, Text Revision (https://psychiatryonline.org/pb-assets/dsm/update/DSM-5-TR_Neurocognitive-Disorders-Supplement_2022_APA_Publishing.pdf). MCI is increasingly recognized as a major healthcare issue because of its association with significant morbidity, including the development of AD dementia. MCI is a heterogeneous clinical syndrome, and new criteria for MCI due to AD are intended to help increase the accuracy of AD diagnosis in the predementia stage [9]. In the Japanese Alzheimer's disease neuroimaging initiative (J-ADNI) study [10], MCI patients (234 patients enrolled, 232 at baseline) progressed to dementia in 12 months at a rate of 28.8% per year, more rapidly than in the ADNI study in the United States (20.2%). The significantly higher rate of conversion to dementia in the J-ADNI persisted until 18 months (39.7% in the J-ADNI vs. 27.4% in the ADNI), whereas the rate in the ADNI caught up after 24 months, reaching 45.1% in the J-ADNI and 40.9% in the ADNI at 36 months. These conversion rates are consistent with those described previously (7–20%) [7–9], with any variations likely to be related to the different measures used to define MCI across studies.

With the recent progress in biomarkers, Jack et al. [11] proposed an unbiased descriptive taxonomy for AD biomarkers. They proposed the "A/T/N" system, dividing the seven major AD biomarkers into three binary categories based on the nature of the pathophysiology of each measure: "A" for amyloid-β biomarker: amyloid positron emission tomography (PET) or cerebrospinal fluid (CSF) amyloid-β42, "T" for tau biomarker: CSF phosphorylated tau or tau PET, and "N" for biomarkers of neurodegeneration or neuronal injury: ^{18}F-fluorodeoxyglucose PET, structural magnetic resonance imaging (MRI), or CSF total tau. Each biomarker category is rated as positive or negative. This is a descriptive system for classifying multidomain biomarker findings at the individual level in a format that is easy to understand and use, as it is agnostic to the temporal sequence of mechanisms underlying AD pathogenesis. Meanwhile, lecanemab, an anti-amyloid-β monoclonal antibody, was approved in June 2023 in the United States and September 2023 in Japan [12] based on the slowing of clinical progression in patients with MCI and early AD. In light of this situation, the role of amyloid PET is becoming increasingly important. In this chapter, we present the current status of amyloid PET imaging approaches in clinical practice and research for AD with an emphasis on the quantification of amyloid deposition.

2 PET Radiopharmaceutical

N-methyl-[^{11}C]2-(4'-methyl-aminophenyl)-6-hydroxybenzotria-zole, or simply Pittsburgh Compound-B (PiB) [13], one of a number of compounds developed for imaging amyloid, is a derivative of the amyloid-binding dye thioflavin-T, the most extensively validated tracer. It binds to aggregated, fibrillar amyloid-β deposits, such as those found in the cerebral cortex and striatum, but not to amorphous amyloid-β deposits, such as those found mostly in the cerebellum. The first ^{11}C-PiB PET study in humans [14] was performed in patients with mild AD, and the uptake pattern was consistent with the deposition of amyloid-β plaques reported in postmortem studies of AD brains. Postmortem studies of patients with high deposition of ^{11}C-PiB during life showed a high correlation between in vivo ^{11}C-PiB accumulation and in vitro measures of amyloid-β pathology [15, 16]. However, due to the 20-min half-life of ^{11}C, ^{11}C-PiB can only be used in PET centers with on-site cyclotron and radiopharmaceutical production facilities.

^{18}F is a more suitable radionuclide for widespread clinical use because its longer half-life of 110 min allows delivery from radiopharmaceutical companies to multiple PET centers. Three ^{18}F-labeled tracers have been developed whose automated synthesizers are approved by pharmaceutical and medical device law in Japan but not reimbursed. Flutemetamol [17] (Vizamyl, GE Healthcare; synthesized by FASTlab, GE Healthcare) is a close structural analog of ^{11}C-PiB, whereas florbetapir [18] (Amyvid, Eli Lilly; synthesized by NEPTIS, Eli Lilly, and MPS200Aβ, Sumitomo Heavy Industries) and florbetaben [19] (Neuraceq, Life Molecular Imaging; synthesized by Synthera, SCETI) are derived from stilbene. Marketing authorizations were granted by the pharmaceutical and medical devices agency for flutemetamol (Vizamyl, Nihon Medi-Physics) and florbetapir (Amyvid, PDRadiopharma Inc.), which are not reimbursed either in Japan. These ^{18}F-labeled tracers are approved for "visualization of amyloid-beta plaques in the brain of patients with mild cognitive impairment or suspected dementia due to Alzheimer's disease." Although ^{11}C-PiB PET images in AD usually show greater binding to gray matter than to white matter, this is not the case for these ^{18}F-labeled tracers, which will be the mainstream of clinical practice shortly. These ^{18}F-labeled tracers have high nonspecific white matter uptake and show a characteristic white matter pattern in scans of amyloid-negative individuals [20]. This negative pattern resembles the smoothed white matter image segmented from whole-brain T1-weighted

MRI. In AD, the ^{18}F-labeled tracers frequently show loss of the gray matter–white matter boundaries, and these consequent losses of normal white matter pattern, as the predominant evidence of cortical amyloid plaque, less often show the intense binding in the cortical ribbon typical of a positive ^{11}C-PiB scan.

On the other hand, ^{18}F-NAV4694 [21] has a close structural resemblance to ^{11}C-PiB and displays imaging characteristics nearly identical to those of ^{11}C-PiB [22]. The low white matter and high cortical binding in AD make this tracer suitable for both clinical and research use, especially for detecting small amounts of amyloid deposition, although this tracer is not approved by pharmaceutical and medical device law in Japan.

3 Clinical Impact of Amyloid PET

The diagnostic impact and clinical utility of amyloid PET imaging have been investigated. Although negative findings on amyloid PET can almost completely rule out AD [23], the prevalence of positive scans in one meta-analysis was 88% in patients with AD, 51% in patients with dementia with Lewy bodies, 30% in patients with cerebrovascular disease, 12% in patients with frontotemporal dementia, 38% in patients with corticobasal degeneration, and 24% in healthy elderly individuals serving as controls [24]. In other meta-analyses of the clinical impact of amyloid PET, the overall rates of change in diagnosis and patient management after amyloid PET varied widely across studies from 19% to 55% [25–32] and from 37% to 87% [27–34], respectively. Grundman et al. [25] found changes in diagnosis after disclosure of the PET results using ^{18}F-florbetapir in 125 of 229 patients (54.6%) as well as in 37.2% of patients with a prescan AD diagnosis and 61.9% of those with a prescan non-AD diagnosis. Another multicenter study using ^{18}F-florbetapir [29] demonstrated changes in diagnosis after disclosure of the PET results in 62 of 228 patients (27.2%) as well as in 27.9% of patients with a prescan AD diagnosis and 25.4% of those with a prescan non-AD diagnosis. In the large-scale IDEAS study [31], amyloid PET results led to changes in the etiologic diagnosis from AD to non-AD in 2869 of 11,409 patients (25.1%) and from non-AD to AD in 1201 of 11,409 patients (10.5%). The higher rate of change in prescan AD may be because a negative amyloid PET can rule out AD. Furthermore, focusing on diagnostic details, after amyloid PET, prescan AD and MCI diagnoses decreased, while postscan diagnoses were further subdivided, increasing from 13 to 21 diagnoses in our multicenter study [32]. This subdivision suggests that amyloid PET contributes to a more reliable diagnosis.

Such changes in diagnosis led to changes in management in 42% of the patients, primarily in medication and care planning in our multicenter study [32]. The most commonly reported change in patient management due to amyloid PET results is a change in medication, ranging from 20% to 60% of cases [28, 30, 31, 34, 35]. The care plans were changed in 10.9% [27] and 46.4% [28] of cases. Amyloid PET results also led to changes in prescription dosage and additional diagnostic tests [26, 33, 34]. Thus, amyloid PET had a considerable impact on diagnostic changes and patient management by improving diagnostic reliability.

4   Visual Interpretation of Amyloid PET

In clinical practice for dementia, amyloid PET increases the certainty of the diagnosis of AD and non-AD [35]. The binary classification of positive and negative amyloid PET findings is based on visual interpretation by trained readers. Visual interpretation guidelines are provided for ^{18}F-florbetapir, ^{18}F-flutemetamol, and ^{18}F-florbetaben.

A review of ^{18}F-florbetapir (https://www.ema.europa.eu/en/documents/product-information/amyvid-epar-product-information_en.pdf) includes all transaxial slices of the brain using a black and white scale with the maximum intensity of the scale set to the maximum intensity of all brain voxels. Interpretation of the images as negative or positive is made by visually comparing the activity in cortical gray matter with activity in adjacent white matter. Negative scans show more radioactivity in white matter than in gray matter, creating a clear gray–white contrast. In contrast, positive scans show cortical areas with a reduction or loss of the normally distinct gray–white contrast. These scans have one or more areas with increased cortical gray matter signal, which results in reduced or absent gray–white contrast. Specifically, a positive scan will have either (a) two or more brain areas (each larger than a single cortical gyrus) in which there is reduced or absent gray–white contrast or (b) one or more areas in which gray matter radioactivity is intense and exceeds radioactivity in adjacent white matter.

A review of ^{18}F-flutemetamol (https://www.ema.europa.eu/en/documents/product-information/vizamyl-epar-product-information_en.pdf) scales images to 90% of the pons activity using a Sokoloff, rainbow, or spectrum color scaling. A negative scan has more radioactivity in the white matter than in the gray matter, creating a clear gray/white matter contrast. In contrast, a positive scan has one or more regions out of five regions comprising the posterior cingulate gyrus and precuneus, frontal cortex, lateral temporal cortex, parietal cortex, and striatum, in which gray matter radioactivity is as intense as, or exceeds, the intensity in the adjacent white matter.

A review of ^{18}F-florbetaben (https://www.ema.europa.eu/en/documents/product-information/neuraceq-epar-product-information_en.pdf) includes all transaxial slices of the brain using a white and black scale. Each of these brain regions, namely, the lateral temporal, frontal, posterior cingulate, precuneus, and parietal lobes, should be systematically visually assessed and scored according to the regional cortical tracer uptake score: 1 (not tracer uptake), 2 (moderate tracer uptake), and 3 (pronounced tracer uptake). The negative scan is defined as a regional cortical tracer uptake score of 1 in each of the four brain regions (lateral temporal lobes, frontal lobes, posterior cingulate/precuneus, parietal lobes). A positive scan with moderate amyloid-β deposition is defined as a regional cortical tracer uptake score of 2 in any or all of the four brain regions and no score of 3 in these four brain regions. A positive scan with pronounced amyloid-β deposition is defined as a regional cortical tracer uptake score of 3 at least in one of four brain regions.

Several spatial and temporal orderings of amyloid positivity using PET have been reported in AD. Huang et al. [36] found early amyloid accumulation in the precuneus, frontal cortex, and posterior cingulate gyrus and later deposition in the parietal, occipital, and lateral temporal cortex. Other studies [37–40] have reported a similar order of amyloid positivity in cortical regions, with early deposition in the medial frontal cortex and precuneus and posterior cingulate gyrus followed by accumulation in the lateral temporal cortex and parietal cortex. Deposition in the striatum has been reported to follow cortical deposition [41].

Familial AD with different autosomal dominant mutations has higher striatal tracer uptake and slightly lower cortical tracer uptake than sporadic AD. Striatal amyloid deposition may be an early feature of familial AD [42, 43]. The cerebellum also shows slightly higher uptake in familial than sporadic AD. Similar to autosomal dominant AD, adults with Down syndrome carry three copies of the amyloid precursor protein gene on chromosome 21 and are genetically predisposed to increased amyloid deposition. In individuals with Down syndrome, amyloid tracer uptake becomes evident around age 40, with an age-dependent increase in spatial extent [44]. The first brain region for increased tracer uptake is the striatum, followed by the rostral prefrontal–cingulo–parietal regions.

In contrast, the physiological increase of amyloid deposition associated with advancing age has been reported to be restricted to the temporal neocortex in cognitively normal individuals [45]. The negative amyloid PET images, AD-related positive images, and aging-related positive images are shown in Fig. 1.

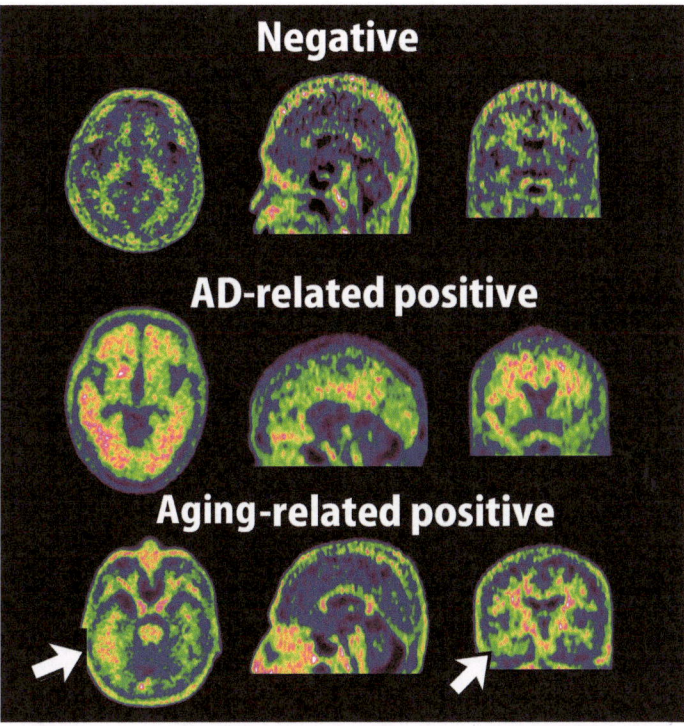

Fig. 1 Amyloid PET images using ^{18}F-NAV4694. Amyloid negative, Alzheimer's disease (AD)-related positive, and aging-related positive cases. The negative case shows no deposition in the cerebral cortex. AD-related amyloid disposition is observed in the frontal, temporal, parietal, occipital, posterior cingulate gyrus, precuneus, and striatum. Aging-related deposition is observed in the right temporal cortex (arrows)

5   Quantification of Amyloid PET

5.1   Centiloid Project

Because raters have their own experience and potential internal criteria, equivocal findings are inevitable and lead to interrater variability in visual interpretation [46, 47]. In our previous multi-center study using ^{18}F-flutemetamol [48], there was disagreement in 9% of cases between two readers. This disagreement rate increases as the number of readers increases. Another multicenter study using ^{18}F-florbetapir [32] showed disagreement in 21% of cases between four readers. Equivocal findings should be avoided when determining the indication for the disease-modifying drugs already approved in Japan. Accordingly, quantitative analysis has been proposed as an adjunct to visual interpretation [49].

To aid the visual interpretation, the standardized uptake value ratio (SUVR) is widely applied to amyloid PET for quantitative measurements of amyloid deposition in the brain. However, SUVR values vary depending on the target and reference regions used as well as the particular amyloid tracer used. This variability can be resolved through a Centiloid [50] scaling process, which standardizes the quantitative amyloid imaging measures by standardizing the outcome of each analytical method and PET ligand. The Centiloid scale named "Centiloids" is anchored at 0 and 100, with 0 Centiloids reflecting a definitively amyloid-negative brain (originally calculated as the average value of a group of healthy subjects below the age of 45) and 100 Centiloids reflecting the average signal observed in patients with typical mild or moderate AD dementia. This harmonized method (originally anchored to ^{11}C-PiB, but now applicable to ^{18}F-labeled tracers after calibration) can facilitate the direct comparison of results across institutions, even when different analytical methods or tracers are used, and may enable cutoffs for amyloid positivity to be clearly defined (Fig. 2).

Several studies have reported threshold values on the Centiloids of amyloid positivity. Salvadó et al. [51] proposed an optimal cutoff value of 12 Centiloids by comparing with amyloid-β 42 levels in CSF. They also proposed that a cutoff value above 29 Centiloids, by comparison with the phosphorylated tau/amyloid-β 42 ratios in CSF, indicates the establishment of AD pathology. On the other hand, Jack et al. [52] estimated a single cutoff value of 19 Centiloids based on a longitudinal study. In a comparative study of antemortem PET and postmortem neuropathology, a threshold of 20.1 Centiloids was most accurate for detecting moderate-to-high-frequent neuritic plaques in CERAD, while a 10 Centiloids or less was optimal for excluding neuritic plaques [53]. La Joie et al. [54] also reported a threshold of 12.2 Centiloids to detect moderate-to-frequent neuritic plaques and a threshold of 24.4 Centiloids for intermediate-to-high AD neuropathological changes. Positive visual interpretation has been reported to be very consistent for 26 Centiloids and higher [32, 53]. On the other hand, a Centiloids of 12–30 is often an equivocal finding in visual interpretation and is referred to as the gray zone [55, 56]. These studies suggest that certainty of visual decipherment of amyloid positivity may be obtained at 26–29 Centiloids or higher.

5.2 Centiloids Calculation

To determine the Centiloids, it is necessary to follow the method put forward by the Global Alzheimer's Association Interactive Network (GAAIN, http://www.gaain.org/centiloid-project). In addition, the processing pipeline for Centiloids calculation needs to be validated using the GAAIN dataset of ^{11}C-PiB PET images for 34 young control individuals and 45 typical AD patients downloaded from the GAAIN website. The Centiloids calculated by the investigator's pipeline should be compared to the Centiloids published on the GAAIN website. For validation, as defined by Klunk et al. [50], the slope should be between 0.98 and 1.02, the intercept between −2 and +2 for a linear regression equation, and the R^2 correlation coefficient should exceed 0.98.

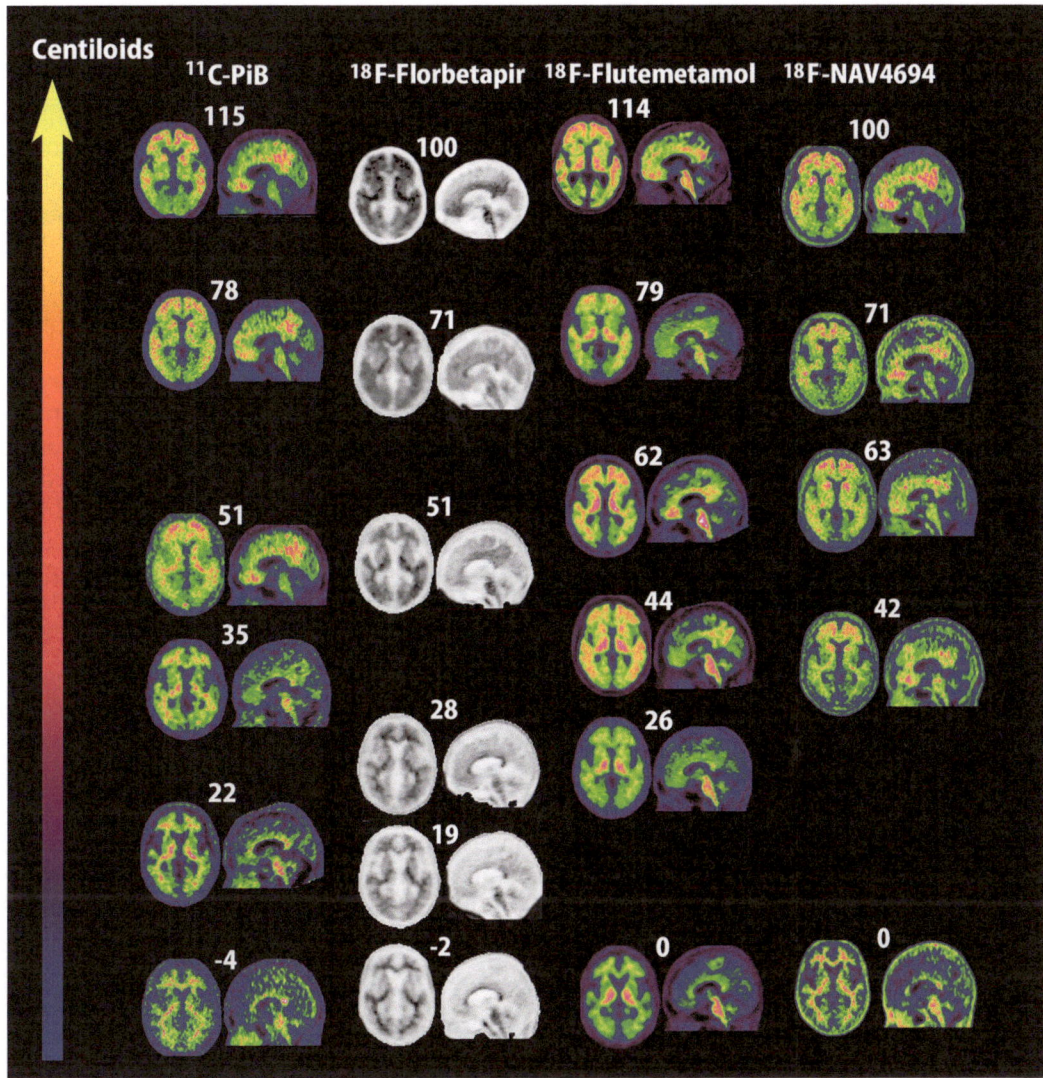

Fig. 2 Amyloid positron emission tomography (PET) images with different tracers scaled by Centiloids. Different PET tracers can be used to measure the same level of amyloid deposition using the Centiloid scale. ^{11}C-PiB, ^{18}F-florbetapir, ^{18}F-flutemetamol, and ^{18}F-NAV4694 amyloid PET images are shown with Centiloids

5.2.1 Standard MRI-Guided Method

In the standard pipeline of Centiloids calculation, the subject's three-dimensional T1-weighted MRI is first reoriented by setting the origin around the anterior commissure and coregistered to the Montreal Neurological Institute (MNI) template (avg152T1.nii). The subject amyloid PET is then reoriented by setting the origin around the anterior commissure and coregistered to the coregistered subject MRI. Next, the coregistered subject MRI was warped into MNI space using unified segmentation [57]. The deformation field parameters in this warping are applied to the coregistered subject PET images for anatomic standardization into MNI space. These translations are performed using the Statistical Parametric Mapping (SPM) 8 or 12 software (https://www.fil.ion.ucl.ac.uk/spm). Global SUVR is calculated from standardized subject PET counts in four reference regions—the whole cerebellum, cerebellar gray matter, pons, and whole cerebellum plus brainstem—and the global cortical target region. Finally, the SUVR is converted to Centiloids using the direct conversion equations for each PET tracer, as described in previous reports [50, 58–61]. The whole cerebellum is reported to be the best standard reference, with the smallest variance and largest effect size [50].

5.2.2   Low-Dose CT-Guided Method

If 3D T1-weighted images are not available, low-dose CT images taken with a PET/CT scanner at a current 40–80 mA may be an alternative. Pressoto et al. [62] reported that anatomic standardization of amyloid PET using ^{18}F-Florbetaben can be performed with low-dose CT used to correct PET attenuation in a PET/CT scanner. They report that the difference between the SUVR measured by anatomic standardization using MRI and low-dose CT is only 0.01 ± 0.03 (mean $\pm$ standard deviation) with Pearson $r = 0.994$, which is negligible. In our comparative study using ^{18}F-flutemetamol, anatomic standardization using low-dose CT resulted in a slight but significant underestimation of -0.01 ± 0.02 and -1.7 ± 2.4 for SUVR and Centiloids, respectively, with Pearson $r = 0.998$ [63]. Similarly, another comparative study using ^{18}F-florbetapir anatomic standardization using low-dose CT yielded a slight but significant underestimation of -2.1 ± 4.2 for Centiloids with Pearson $r = 0.995$ [64]. This slight underestimation of SUVR and Centiloids in the low-dose CT-guided method may be due to slight differences in PET counts in the frontal cortex or cerebellar cortex compared to the MRI-guided method [63, 64]. Kim et al. [65] also reported the feasibility of low-dose CT in Centiloids calculation using ^{18}F-florbetaben and ^{18}F-flutemetamol. The main advantage of the use of low-dose CT is the simultaneous acquisition of PET and CT using a PET/CT scanner. This method would apply to subjects who are unable to undergo MRI for various reasons.

5.2.3   PET-Alone Method

In contrast, several studies [66–74] have reported anatomic standardization using the amyloid PET template alone without MRI. Among these studies, those using an adaptive template generated from a linear combination of an amyloid-negative and amyloid-positive template with a weight-optimized algorithm have yielded Centiloids comparable to those obtained with the standard MRI-guided method [69, 70, 72, 74]. However, a correlation analysis of this PET-alone method with the standard MRI-guided method showed that the Centiloids deviated slightly from the validation criteria for ¹⁸F-florbetapir from the Alzheimer's Disease Neuroimaging Initiative dataset (slope, 1.028; intercept, −4.302; R^2, 0.974) [72]. Errors in anatomic standardization may occur when the distribution of amyloid PET differs from that of the PET template, such as when there is a marked left-right difference in amyloid PET deposition. In such cases, anatomic standardization using coregistered structural images may be more accurate than the PET-alone method.

5.3 Quantification Software Packages

Several software packages have been developed for amyloid PET quantification; SUVR and Centiloids help detect amyloid positivity by aiding visual assessment [75]. These software packages are also expected to be applied to observe the longitudinal observation of amyloid deposition and to determine the efficacy of disease-modifying drugs. In the evaluation of amyloid positivity, it is useful to demonstrate not only a global Centiloids but also a local Z-score map of amyloid PET in patients compared to a negative database of healthy individuals below 50 years old, and such software packages have also been developed [64, 76, 77]. This Z-score map may be extremely useful as an aid of visual interpretation (Fig. 3). Healthy

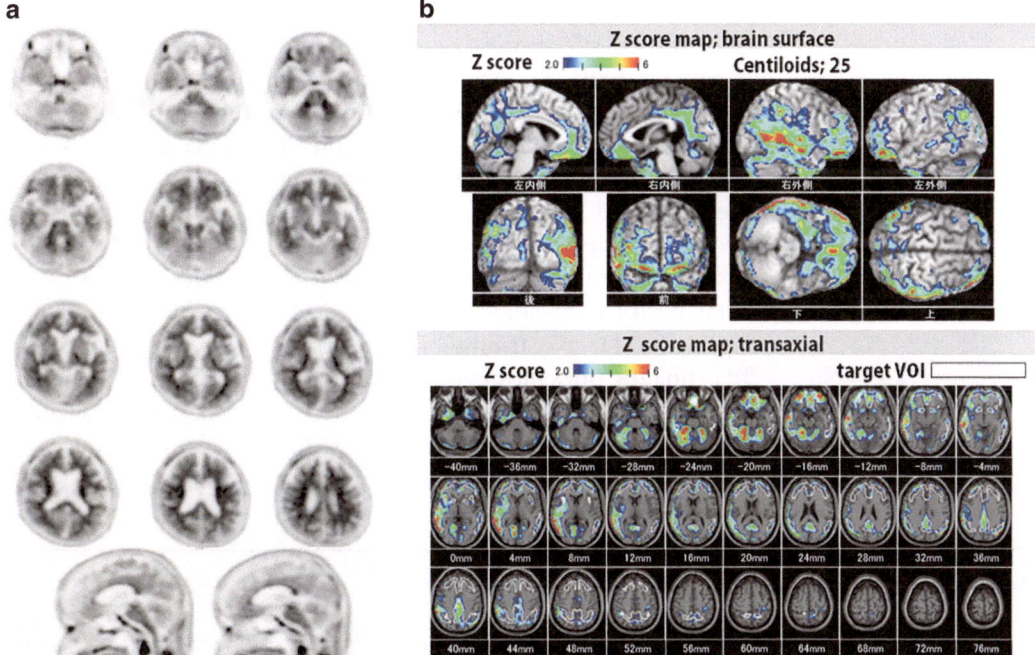

Fig. 3 Z-score mapping to aid in visual interpretation. (**a**) Transaxial and sagittal PET images using ¹⁸F-florbetapir. Positive and negative results varied from reader to reader. (**b**) Z-score mapping compared to a negative normal database reveals statistically significant amyloid deposition above the Z-score of 2 in the frontal, temporal, and posterior cingulate gyrus and precuneus. Centiloids is 25. Amyquant software [77] was used for Centiloids calculation and Z-score mapping

individuals under 50 years of age are preferred for a negative database because it is unlikely that anyone in this age group has been diagnosed with preclinical AD [78].

5.4 Applications of Artificial Intelligence to Amyloid PET

Recent developments in artificial intelligence are opening new vistas in amyloid PET imaging technology, diagnosis, and analysis. Deep learning methods synthesize diagnostic-quality PET images from very low-dose PET images at 1/50 of the standard dose [79]. This deep learning method can also reduce PET scan time from the standard level of 20 to 5 min [80] or even 2 min [81], while still maintaining acceptable image quality for the correct diagnosis of amyloid deposition. This reduction may be an attractive solution to the clinical need for improved PET imaging workflow.

Deep learning methods may support the visual interpretation of amyloid PET images. Deep learning methods increased the kappa coefficient of inter-reader agreement for visual reading from 0.46 to 0.76 [82]. Such an increase in inter-reader agreement with deep learning methods is observed to be more pronounced for visually equivocal scans [83]. Among deep learning methods, a convolutional neural network is a type of artificial neural network, which is widely used for image recognition and classification. A sagittal two-dimensional convolutional neural network provided a successful classification of negative and positive amyloid PET scans in patients with subjective cognitive decline [84]. The convolutional neural networks replicated the expert classification and quantitative measure of amyloid deposition [85]. This method has the potential to replace or simplify existing clinical routines and facilitate fast and accurate classification well suited for a high-throughput pipeline.

Deep learning methods have also been reported to be useful when performing anatomic standardization to quantify amyloid deposition. Deep learning–based spatial normalization of amyloid PET images does not require MRI or CT images of the same patient [86, 87]. Deep learning methods can also improve accuracy in quantifying amyloid deposition by removing undesirable non-specific uptake [88].

6 Conclusions

Amyloid PET is now largely established as an imaging tool for the diagnosis of cognitively impaired patients and has great potential for in vivo identification of amyloid pathology and biological staging of neurodegenerative diseases. In addition, it has helped in distinguishing between AD and other dementia subtypes with overlapping clinical phenotypes and identifying atypical Alzheimer's disease variants. Amyloid PET has contributed significantly to the development of disease-modifying therapies for AD. The aducanumab trial demonstrated that the "amyloid hypothesis" may be true in some patients [89]. A phase III trial of lecanemab in early symptomatic AD patients with PET evidence of amyloid deposition showed a 59.1 Centiloids reduction in amyloid plaque levels and a milder increase in Clinical Dementia Rating-Sum of Boxes at 18 months as compared to placebo [90]. Similarly, a phase II trial of donanemab in early symptomatic AD patients with PET evidence of amyloid deposition reduced amyloid plaque levels by 85.06 Centiloids and lessened the decline of Integrated Alzheimer's Disease Rating Scale at week 76 as compared to placebo [91]. These findings indicate that targeting a reduction in amyloid-β may prevent cognitive decline. The clinical importance of amyloid PET will undoubtedly grow in the future. Quantitative assessment of amyloid deposition is expected to make visual evaluation more accurate. Furthermore, the clinical importance of amyloid PET quantification will further increase given the use of the amyloid PET Centiloids in an ongoing clinical trial of lecanemab in cognitively unimpaired individuals [92]. Furthermore, a more accurate diagnosis of AD is becoming possible when combined with PET tracers that accumulate in the tau protein characteristic of AD [93].

Acknowledgments

We thank all clinicians and imaging technicians who contributed to this study.

References

1. Alzheimer's Association (2023) Alzheimer's disease facts and figures. Available from: https://www.alz.org/media/documents/alzheimers-facts-and-figures.pdf

2. Alzheimer's disease international (2023) World Alzheimer report 2023. Available from: https://www.alzint.org/u/World-Alzheimer-Report-2023.pdf

3. McKhann GM, Knopman DS, Chertkow H et al (2011) The diagnosis of dementia due to Alzheimer's disease: recommendations from the National Institute on Aging-Alzheimer's Association workgroups on diagnostic guidelines for Alzheimer's disease. Alzheimers Dement 7:263–269. https://doi.org/10.1016/j.jalz.2011.03.005

4. Dubois B, Feldman HH, Jacova C et al (2014) Advancing research diagnostic criteria for Alzheimer's disease: the IWG-2 criteria. Lancet Neurol 13:614–629. https://doi.org/10.1016/S1474-4422(14)70090-0

5. Jack CR Jr, Albert MS, Knopman DS et al (2011) Introduction to the recommendations from the National Institute on Aging-Alzheimer's Association workgroups on diagnostic guidelines for Alzheimer's disease. Alzheimers Dement 7:257–262. https://doi.org/10.1016/j.jalz.2011.03.004

6. Sperling RA, Aisen PS, Beckett LA et al (2011) Toward defining the preclinical stages of Alzheimer's disease: recommendations from the National Institute on Aging-Alzheimer's Association workgroups on diagnostic guidelines for Alzheimer's disease. Alzheimers Dement 7:280–292. https://doi.org/10.1016/j.jalz.2011.03.003

7. Petersen RC (2004) Mild cognitive impairment as a diagnostic entity. J Intern Med 256:183–194. https://doi.org/10.1111/j.1365-2796.2004.01388.x

8. Gauthier S, Reisberg B, Zaudig M et al (2006) Mild cognitive impairment. Lancet 367:1262–

1270. https://doi.org/10.1016/S0140-6736(06)68542-5

9. Albert MS, DeKosky ST, Dickson D et al (2011) The diagnosis of mild cognitive impairment due to Alzheimer's disease: recommendations from the National Institute on Aging-Alzheimer's Association workgroups on diagnostic guidelines for Alzheimer's disease. Alzheimers Dement 7:270–279. https://doi.org/10.1016/j.jalz.2011.03.008

10. Iwatsubo T, Iwata A, Suzuki K et al (2018) Japanese and North American Alzheimer's disease neuroimaging initiative studies: harmonization for international trials. Alzheimers Dement 14:1077–1087. https://doi.org/10.1016/j.jalz.2018.03.009

11. Jack CR Jr, Bennett DA, Blennow K et al (2016) A/T/N: an unbiased descriptive classification scheme for Alzheimer disease biomarkers. Neurology 87:539–547. https://doi.org/10.1212/WNL.0000000000002923

12. Boxer AL, Sperling R (2023) Accelerating Alzheimer's therapeutic development: the past and future of clinical trials. Cell S0092-8674(23):01077-2. https://doi.org/10.1016/j.cell.2023.09.023

13. Mathis CA, Wang Y, Holt DP et al (2003) Synthesis and evaluation of 11C-labeled 6-substituted 2-arylbenzothiazoles as amyloid imaging agents. J Med Chem 46:2740–2754. https://doi.org/10.1021/jm030026b

14. Klunk WE, Engler H, Nordberg A et al (2004) Imaging brain amyloid in Alzheimer's disease with Pittsburgh compound-B. Ann Neurol 55:306–319. https://doi.org/10.1002/ana.20009

15. Bacskai BJ, Frosch MP, Freeman SH et al (2007) Molecular imaging with Pittsburgh compound B confirmed at autopsy: a case report. Arch Neurol 64:431–434. https://doi.org/10.1001/archneur.64.3.431

16. Ikonomovic MD, Klunk WE, Abrahamson EE (2008) Post-mortem correlates of in vivo PiB-PET amyloid imaging in a typical case of Alzheimer's disease. Brain 131(Pt 6):1630–1645. https://doi.org/10.1093/brain/awn016

17. Nelissen N, Van Laere K, Thurfjell L et al (2009) Phase 1 study of the Pittsburgh compound B derivative 18F-flutemetamol in healthy volunteers and patients with probable Alzheimer disease. J Nucl Med 50:1251–1259. https://doi.org/10.2967/jnumed.109.063305

18. Wong DF, Rosenberg PB, Zhou Y et al (2010) In vivo imaging of amyloid deposition in Alzheimer disease using the radioligand [18F]-AV-45 (florbetapir [corrected] F 18). J Nucl Med 51:913–920. https://doi.org/10.2967/jnumed.109.069088

19. Barthel H, Gertz HJ, Dresel S et al (2011) Cerebral amyloid-β PET with florbetaben ([18F]) in patients with Alzheimer's disease and healthy controls: a multicentre phase 2 diagnostic study. Lancet Neurol 10:424–435. https://doi.org/10.1016/S1474-4422(11)70077-1

20. Rowe CC, Villemagne VL (2011) Brain amyloid imaging. J Nucl Med 52:1733–1740. https://doi.org/10.2967/jnumed.110.076315

21. Cselényi Z, Jönhagen ME, Forsberg A et al (2012) Clinical validation of [18F]-AZD4694, an amyloid-β-specific PET radioligand. J Nucl Med 53:415–424. https://doi.org/10.2967/jnumed.111.094029

22. Rowe CC, Pejoska S, Mulligan RS et al (2013) Head-to-head comparison of [11C]-PiB and [18F]-AZD4694 (NAV4694) for β-amyloid imaging in aging and dementia. J Nucl Med 54:880–886. https://doi.org/10.2967/jnumed.112.114785

23. Barthel H, Sabri O (2017) Clinical use and utility of amyloid imaging. J Nucl Med 58:1711–1717. https://doi.org/10.2967/jnumed.116.185017

24. Ossenkoppele R, Jansen WJ, Rabinovici GD et al (2015) Prevalence of amyloid PET positivity in dementia syndromes: a meta-analysis. JAMA 313:1939–1949. https://doi.org/10.1001/jama.2015.4669

25. Grundman M, Pontecorvo MJ, Salloway SP et al (2013) Potential impact of amyloid imaging on diagnosis and intended management in patients with progressive cognitive decline. Alzheimer Dis Assoc Disord 27:4–15. https://doi.org/10.1097/WAD.0b013e318279d02a

26. Weston PS, Paterson RW, Dickson J et al (2016) Diagnosing dementia in the clinical setting: can amyloid PET provide additional value over cerebrospinal fluid? J Alzheimers Dis 54:1297–1302. https://doi.org/10.3233/JAD-160302

27. Zwan MD, Bouwman FH, Konijnenberg E et al (2017) Diagnostic impact of [18F]flutemetamol PET in early-onset dementia. Alzheimers Res Ther 9:2. https://doi.org/10.1186/s13195-016-0228-4

28. Bensaïdane MR, Beauregard JM, Poulin S et al (2016) Clinical utility of amyloid PET imaging in the differential diagnosis of atypical dementias and its impact on caregivers. J Alzheimers Dis 52:1251–1262. https://doi.org/10.3233/JAD-151180

29. Boccardi M, Altomare D, Ferrari C et al (2016) Assessment of the incremental diagnostic value of florbetapir F 18 imaging in patients with cognitive impairment: the incremental diagnostic value of amyloid PET with [18F]-florbetapir (India-FBP) study. JAMA Neurol 73:1417–1424. https://doi.org/10.1001/jamaneurol.2016.3751

30. Carswell CJ, Win Z, Muckle K et al (2018) Clinical utility of amyloid PET imaging with (18)F-florbetapir: a retrospective study of 100 patients. J Neurol Neurosurg Psychiatry 89:294–299. https://doi.org/10.1136/jnnp-2017-316194

31. Rabinovici GD, Gatsonis C, Apgar C et al (2019) Association of amyloid positron emission tomography with subsequent change in clinical management among medicare beneficiaries with mild cognitive impairment or dementia. JAMA 321:1286–1294. https://doi.org/10.1001/jama.2019.2000

32. Matsuda H, Okita K, Motoi Y et al (2022) Clinical impact of amyloid PET using ^{18}F-florbetapir in patients with cognitive impairment and suspected Alzheimer's disease: a multicenter study. Ann Nucl Med 36:1039–1049. https://doi.org/10.1007/s12149-022-01792-y

33. Ceccaldi M, Jonveaux T, Verger A et al (2018) Added value of ^{18}F-florbetaben amyloid PET in the diagnostic workup of most complex patients with dementia in France: a naturalistic study. Alzheimers Dement 14:293–305. https://doi.org/10.1016/j.jalz.2017.09.009

34. Grundman M, Johnson KA, Lu M et al (2016) Effect of amyloid imaging on the diagnosis and management of patients with cognitive decline: impact of appropriate use criteria. Dement Geriatr Cogn Disord 41:80–92. https://doi.org/10.1159/000441139

35. Leuzy A, Savitcheva I, Chiotis K et al (2019) Clinical impact of [^{18}F]flutemetamol PET among memory clinic patients with an unclear diagnosis. Eur J Nucl Med Mol Imaging 46:1276–1286. https://doi.org/10.1007/s00259-019-04297-5

36. Huang KL, Lin KJ, Hsiao IT et al (2013) Regional amyloid deposition in amnestic mild cognitive impairment and Alzheimer's disease evaluated by [18F]AV-45 positron emission tomography in Chinese population. PLoS One 8:e58974. https://doi.org/10.1371/journal.pone.0058974

37. Villeneuve S, Rabinovici GD, Cohn-Sheehy BI et al (2015) Existing Pittsburgh compound-B positron emission tomography thresholds are too high: statistical and pathological evaluation. Brain 138:2020–2033. https://doi.org/10.1093/brain/awv112

38. Cho H, Choi JY, Hwang MS et al (2016) In vivo cortical spreading pattern of tau and amyloid in the Alzheimer disease spectrum. Ann Neurol 80:247–258. https://doi.org/10.1002/ana.24711

39. Palmqvist S, Schöll M, Strandberg O et al (2017) Earliest accumulation of β-amyloid occurs within the default-mode network and concurrently affects brain connectivity. Nat Commun 8:1214. https://doi.org/10.1038/s41467-017-01150-x

40. Mattsson N, Insel PS, Donohue M et al (2015) Independent information from cerebrospinal fluid amyloid-β and florbetapir imaging in Alzheimer's disease. Brain 138:772–783. https://doi.org/10.1093/brain/awu367

41. Hanseeuw BJ, Betensky RA, Mormino EC et al (2018) PET staging of amyloidosis using striatum. Alzheimers Dement 14:1281–1292. https://doi.org/10.1016/j.jalz.2018.04.011

42. Klunk WE, Price JC, Mathis CA et al (2007) Amyloid deposition begins in the striatum of presenilin-1 mutation carriers from two unrelated pedigrees. J Neurosci 27:6174–6184. https://doi.org/10.1523/JNEUROSCI.0730-07.2007

43. Villemagne VL, Ataka S, Mizuno T et al (2009) High striatal amyloid beta-peptide deposition across different autosomal Alzheimer disease mutation types. Arch Neurol 66:1537–1544. https://doi.org/10.1001/archneurol.2009.285

44. Annus T, Wilson LR, Hong YT et al (2016) The pattern of amyloid accumulation in the brains of adults with Down syndrome. Alzheimers Dement 12:538–545. https://doi.org/10.1016/j.jalz.2015.07.490

45. Gonneaud J, Arenaza-Urquijo EM, Mézenge F et al (2017) Increased florbetapir binding in the temporal neocortex from age 20 to 60 years. Neurology 89:2438–2446. https://doi.org/10.1212/WNL.0000000000004733

46. Hosokawa C, Ishii K, Kimura Y et al (2015) Performance of ^{11}C-Pittsburgh compound B PET binding potential images in the detection of amyloid deposits on equivocal static images. J Nucl Med 56:1910–1915. https://doi.org/10.2967/jnumed.115.156414

47. Yamane T, Ishii K, Sakata M et al (2017) Inter-rater variability of visual interpretation and comparison with quantitative evaluation of ^{11}C-PiB PET amyloid images of the Japanese Alzheimer's disease neuroimaging initiative (J-ADNI) multicenter study. Eur J Nucl Med

Mol Imaging 44:850–857. https://doi.org/
10.1007/s00259-016-3591-2

48. Matsuda H, Ito K, Ishii K et al (2021) Quantitative evaluation of ^{18}F-flutemetamol PET in patients with cognitive impairment and suspected Alzheimer's disease: a multicenter study. Front Neurol 11:578753. https://doi.org/10.3389/fneur.2020.578753

49. Collij LE, Konijnenberg E, Reimand J et al (2019) Assessing amyloid pathology in cognitively normal subjects using ^{18}F-flutemetamol PET: comparing visual reads and quantitative methods. J Nucl Med 60:541–547. https://doi.org/10.2967/jnumed.118.211532

50. Klunk WE, Koeppe RA, Price JC et al (2015) The Centiloid project: standardizing quantitative amyloid plaque estimation by PET. Alzheimers Dement 11:1–15.e154. https://doi.org/10.1016/j.jalz.2014.07.003

51. Salvadó G, Molinuevo JL, Brugulat-Serrat A et al (2019) Centiloid cut-off values for optimal agreement between PET and CSF core AD biomarkers. Alzheimers Res Ther 11:27. https://doi.org/10.1186/s13195-019-0478-z

52. Jack CR Jr, Wiste HJ, Weigand SD et al (2017) Defining imaging biomarker cut points for brain aging and Alzheimer's disease. Alzheimers Dement 13:205–216. https://doi.org/10.1016/j.jalz.2016.08.005

53. Amadoru S, Doré V, McLean CA et al (2020) Comparison of amyloid PET measured in Centiloid units with neuropathological findings in Alzheimer's disease. Alzheimers Res Ther 12:22. https://doi.org/10.1186/s13195-020-00587-5

54. La Joie R, Ayakta N, Seeley WW et al (2019) Multisite study of the relationships between antemortem [11C]PIB-PET Centiloid values and postmortem measures of Alzheimer's disease neuropathology. Alzheimers Dement 15:205–216. https://doi.org/10.1016/j.jalz.2018.09.001

55. Pemberton HG, Collij LE, Heeman F et al (2022) Quantification of amyloid PET for future clinical use: a state-of-the-art review. Eur J Nucl Med Mol Imaging 49:3508–3528. https://doi.org/10.1007/s00259-022-05784-y

56. Milà-Alomà M, Salvadó G, Shekari M et al (2021) Comparative analysis of different definitions of amyloid-β positivity to detect early downstream pathophysiological alterations in preclinical Alzheimer. J Prev Alzheimers Dis 8:68–77. https://doi.org/10.14283/jpad.2020.51

57. Ashburner J, Friston KJ (2005) Unified segmentation. NeuroImage 26:839–851. https://doi.org/10.1016/j.neuroimage.2005.02.018

58. Navitsky M, Joshi AD, Kennedy I et al (2018) Standardization of amyloid quantitation with florbetapir standardized uptake value ratios to the Centiloid scale. Alzheimers Dement 14:1565–1571. https://doi.org/10.1016/j.jalz.2018.06.1353

59. Battle MR, Pillay LC, Lowe VJ et al (2018) Centiloid scaling for quantification of brain amyloid with [18F]flutemetamol using multiple processing methods. EJNMMI Res 8:107. https://doi.org/10.1186/s13550-018-0456-7

60. Rowe CC, Jones G, Doré V et al (2016) Standardized expression of ^{18}F-NAV4694 and ^{11}C-PiB β-amyloid PET results with the Centiloid scale. J Nucl Med 57:1233–1237. https://doi.org/10.2967/jnumed.115.171595

61. Rowe CC, Doré V, Jones G et al (2017) ^{18}F-Florbetaben PET beta-amyloid binding expressed in Centiloids. Eur J Nucl Med Mol Imaging 44:2053–2059. https://doi.org/10.1007/s00259-017-3749-6

62. Presotto L, Iaccarino L, Sala A et al (2018) Low-dose CT for the spatial normalization of PET images: a validation procedure for amyloid-PET semi-quantification. Neuroimage Clin 220:153–160. https://doi.org/10.1016/j.nicl.2018.07.013

63. Matsuda H, Yamao T, Shakado M et al (2021) Amyloid PET quantification using low-dose CT-guided anatomic standardization. EJNMMI Res 11:125. https://doi.org/10.1186/s13550-021-00867-7

64. Matsuda H, Soma T, Okita K et al (2023) Development of software for measuring brain amyloid accumulation using ^{18}F-florbetapir PET and calculating global Centiloid scale and regional Z-score values. Brain Behav 13:e3092. https://doi.org/10.1002/brb3.3092

65. Kim SJ, Ham H, Park YH et al (2022) Development and clinical validation of CT-based regional modified Centiloid method for amyloid PET. Alzheimers Res Ther 14:157. https://doi.org/10.1186/s13195-022-01099-0

66. Edison P, Carter SF, Rinne JO et al (2013) Comparison of MRI based and PET template based approaches in the quantitative analysis of amyloid imaging with PIB-PET. NeuroImage 70:423–433. https://doi.org/10.1016/j.neuroimage.2012.12.014

67. Lundqvist R, Lilja J, Thomas BA et al (2013) (2013) implementation and validation of an

adaptive template registration method for 18F-flutemetamol imaging data. J Nucl Med 54:1472–1478. https://doi.org/10.2967/jnumed.112.115006

68. Saint-Aubert L, Nemmi F, Péran P et al (2014) Comparison between PET template-based method and MRI-based method for cortical quantification of florbetapir (AV-45) uptake in vivo. Eur J Nucl Med Mol Imaging 41:836–843. https://doi.org/10.1007/s00259-013-2656-8

69. Bourgeat P, Villemagne VL, Dore V et al (2015) Comparison of MR-less PiB SUVR quantification methods. Neurobiol Aging 36 (Suppl 1):S159–S166. https://doi.org/10.1016/j.neurobiolaging.2014.04.033

70. Bourgeat P, Doré V, Fripp J et al (2018) Implementing the centiloid transformation for ^{11}C-PiB and β-amyloid ^{18}F-PET tracers using CapAIBL. NeuroImage 183:387–393. https://doi.org/10.1016/j.neuroimage.2018.08.044

71. Tsubaki Y, Akamatsu G, Shimokawa N et al (2020) Development and evaluation of an automated quantification tool for amyloid PET images. EJNMMI Phys 7:59. https://doi.org/10.1186/s40658-020-00329-4

72. Fujishima M, Matsuda H (2022) Non-standard pipeline without MRI has replicability in computation of Centiloid scale values for PiB and 18F-labeled amyloid PET tracers. Neuroimage Rep 2:100101. https://doi.org/10.1016/j.ynirp.2022.100101

73. Iaccarino L, La Joie R, Koeppe R et al (2022) rPOP: robust PET-only processing of community acquired heterogeneous amyloid-PET data. NeuroImage 246:118775. https://doi.org/10.1016/j.neuroimage.2021.118775

74. Imabayashi E, Tamamura N, Yamaguchi Y et al (2022) Automated semi-quantitative amyloid PET analysis technique without MR images for Alzheimer's disease. Ann Nucl Med 36:865–875. https://doi.org/10.1007/s12149-022-01769-x

75. Jovalekic A, Roé-Vellvé N, Koglin N et al (2023) Validation of quantitative assessment of florbetaben PET scans as an adjunct to the visual assessment across 15 software methods. Eur J Nucl Med Mol Imaging 50:3276–3289. https://doi.org/10.1007/s00259-023-06279-0

76. Pemberton HG, Buckley C, Battle M et al (2023) Software compatibility analysis for quantitative measures of [^{18}F]flutemetamol amyloid PET burden in mild cognitive impairment. EJNMMI Res 13:48. https://doi.org/10.1186/s13550-023-00994-3

77. Matsuda H, Yamao T (2022) Software development for quantitative analysis of brain amyloid PET. Brain Behav 12:e2499. https://doi.org/10.1002/brb3.2499

78. Zhang K, Mizuma H, Zhang X et al (2021) PET imaging of neural activity, β-amyloid, and tau in normal brain aging. Eur J Nucl Med Mol Imaging 48:3859–3871. https://doi.org/10.1007/s00259-021-05230-5

79. Chen KT, Adeyeri O, Toueg TN et al (2022) Investigating simultaneity for deep learning-enhanced actual ultra-low-dose amyloid PET/MR imaging. AJNR Am J Neuroradiol 43:354–360. https://doi.org/10.3174/ajnr.A7410

80. Peng Z, Ni M, Shan H et al (2021) Feasibility evaluation of PET scan-time reduction for diagnosing amyloid-β levels in Alzheimer's disease patients using a deep-learning-based denoising algorithm. Comput Biol Med 138:104919. https://doi.org/10.1016/j.compbiomed.2021.104919

81. Jeong YJ, Park HS, Jeong JE et al (2021) Restoration of amyloid PET images obtained with short-time data using a generative adversarial networks framework. Sci Rep 11:4825. https://doi.org/10.1038/s41598-021-84358-8

82. Kim JY, Oh D, Sung K et al (2021) Visual interpretation of [^{18}F]Florbetaben PET supported by deep learning-based estimation of amyloid burden. Eur J Nucl Med Mol Imaging 48:1116–1123. https://doi.org/10.1007/s00259-020-05044-x

83. Son HJ, Oh JS, Oh M et al (2020) The clinical feasibility of deep learning-based classification of amyloid PET images in visually equivocal cases. Eur J Nucl Med Mol Imaging 47:332–341. https://doi.org/10.1007/s00259-019-04595-y

84. de Vries BM, Golla SSV, Ebenau J et al (2021) Classification of negative and positive ^{18}F-florbetapir brain PET studies in subjective cognitive decline patients using a convolutional neural network. Eur J Nucl Med Mol Imaging 48:721–728. https://doi.org/10.1007/s00259-020-05006-3

85. Ladefoged CN, Anderberg L, Madsen K et al (2023) Estimation of brain amyloid accumulation using deep learning in clinical [^{11}C]PiB PET imaging. EJNMMI Phys 10:44. https://doi.org/10.1186/s40658-023-00562-7

86. Kang SK, Seo S, Shin SA et al (2018) Adaptive template generation for amyloid PET using a deep learning approach. Hum Brain Mapp 39:3769–3778. https://doi.org/10.1002/hbm.24210

87. Kang SK, Kim D, Shin SA et al (2023) Fast and accurate amyloid brain PET quantification without MRI using deep neural networks. J Nucl Med 64:659–666. https://doi.org/10.2967/jnumed.122.264414

88. Liu H, Nai YH, Saridin F et al (2021) Improved amyloid burden quantification with nonspecific estimates using deep learning. Eur J Nucl Med Mol Imaging 48:1842–1853. https://doi.org/10.1007/s00259-020-05131-z

89. Budd Haeberlein S, Aisen PS, Barkhof F et al (2022) Two randomized phase 3 studies of Aducanumab in early Alzheimer's disease. J Prev Alzheimers Dis 9:197–210. https://doi.org/10.14283/jpad.2022.30

90. van Dyck CH, Swanson CJ, Aisen P (2023) Lecanemab in early Alzheimer's disease. N Engl J Med 388:9–21. https://doi.org/10.1056/NEJMoa2212948

91. Mintun MA, Lo AC, Duggan Evans C (2021) Donanemab in early Alzheimer's disease. N Engl J Med 384:1691–1704. https://doi.org/10.1056/NEJMoa2100708

92. Rafii MS, Sperling RA, Donohue MC et al (2023) The AHEAD 3-45 study: design of a prevention trial for Alzheimer's disease. Alzheimers Dement 19:1227–1233. https://doi.org/10.1002/alz.12748

93. Matsuda H, Yamao T (2023) Tau positron emission tomography in patients with cognitive impairment and suspected Alzheimer's disease. Fukushima J Med Sci 69:85–93. https://doi.org/10.5387/fms.2023-08

Chapter 6

^{18}F-FDG PET for Neurodegenerative Diseases: Principles and Progress

Tanyaluck Thientunyakit, Nantika Wannasoupol,
Chakmeedaj Sethanandha, Siriwan Piyapittayanan, Tossaporn Siriprapa,
Thonnapong Thongpraparn, Weerasak Muangpaisan, and Juri Gelovani

Abstract

The prevalence of neurodegenerative diseases has been increasing in aging society. Recent developments in several biomarkers help understand the underlying principles of neurodegenerative diseases and accelerate remarkable progress in clinical practice and research. Here, we describe a prospective study on the use of ^{18}F-fluorodeoxyglucose positron emission tomography (^{18}F-FDG PET) of the brain as a predictive marker for response to treatment in patients with Alzheimer's disease and compare results with other imaging markers, including amyloid PET, MRI, and the integrated imaging marker using amyloid PET, FDG PET, and MRI. We aim to provide the readers with guidelines on applying both qualitative and semiquantitative and integrating FDG PET data with other imaging markers. The research challenges and possible measures to overcome those challenges are also discussed.

Key words ^{18}F-fluorodeoxyglucose, Positron emission tomography, ^{18}F-FDG PET, Neuroimaging, Neurodegenerative diseases, Alzheimer's disease, Treatment response

1 Introduction

Neurodegenerative diseases, which impair a person's cognition and ability in daily tasks, have been steadily increasing in an aging society and remain one of the most formidable challenges in the field of neuroscience. Alzheimer's disease (AD) is the most common form of neurodegenerative disease [1]. It is known as one of the leading causes of death in the elderly that significantly affects global health [2]. One primary concern for AD is that current pharmacological treatment strategies are limited and costly. Current medications, acetylcholinesterase inhibitors (AchEIs) (such as donepezil, galantamine, and rivastigmine), and noncompetitive N-methyl-D-aspartate (NMDA) receptor antagonists (such as memantine) possibly benefit in mild-to-moderate severity of AD. However, the response rates are low and may not be suitable for

Daichi Sone (ed.), *Molecular Imaging for Brain Diseases*, Neuromethods, vol. 222, https://doi.org/10.1007/978-1-0716-4494-2_6,

patients with other medical conditions, for example, multiple sclerosis, HIV, and chronic epilepsy. Moreover, previous studies found that the response rate is associated with the disease severity and rate of progression [3–6]. Thus, an appropriate selection of patients who would likely respond to such treatment would result in a more cost-effective approach and, therefore, optimal benefit for both patients and physicians in clinical practice.

Recent developments in several biomarkers, such as ^{18}F-fluorodeoxyglucose (^{18}F-FDG) positron emission tomography (PET), have played a pivotal role in understanding the underlying pathophysiological changes in AD, which helps diagnose and monitor neurological and psychiatric disorders in both clinical practice and research. ^{18}F-FDG is a radiolabeled glucose analog used to examine brain glucose metabolism that relies on glucose transporter proteins, particularly GLUT-1 and GLUT-3, to facilitate the transport of FDG into neurons and glial cells by mimicking glucose's structure [7]. Reduced synaptic metabolism of glucose and FDG uptake suggest synaptic vulnerability to AD [8]. FDG PET is a relatively low-cost and widely available imaging technique that has demonstrated clinical usefulness in the diagnosis of neurodegenerative diseases, including AD and mild cognitive impairment (MCI) [9]. FDG PET imaging patterns help diagnose, distinguish, and track diseases because they can show unique metabolic fingerprints linked to various stages of neurodegenerative disorders. Widespread and progressive hypometabolism in the precuneus, posterior cingulate gyrus, temporoparietal, and frontal cortices, which correlate with clinical symptoms and disease severity, is a key AD feature. Another characteristic of AD is the sparing of the sensorimotor cortex and relatively preserved metabolism in the thalamus, posterior fossa, basal ganglia, and visual and anterior cingulate cortices. As neuronal dysfunction and loss are the primary characteristics of dementia, FDG PET provides a dynamic and functional insight into brain health.

FDG PET is often used in conjunction with other neuroimaging techniques, such as MRI, to provide a more comprehensive view of brain proteinopathy, function, and structure, even though it shows promise for tracking disease progression over time and may help select suitable candidates for therapies and interventions to slow or mitigate the cognitive decline associated with AD. Integrating two (or more) imaging modalities offers several methodologic benefits. While FDG PET can detect an early metabolic alteration in the brain, structural MR imaging can distinguish between neurodegenerative dementias and help rule out non-neurodegenerative, potentially curable causes of cognitive impairments. On the other hand, one of the essential hallmarks of AD is cerebral amyloid deposition, which can be verified by amyloid PET. Cerebral amyloid deposition can be identified up to 20 years earlier before patients develop symptomatic dementia. In contrast, neurodegenerative indicators like FDG PET are known to have a higher association and greater specificity with the clinical severity of Alzheimer's disease [10]. In addition to improving diagnostic classification, disease staging, prognostic assessment, and early diagnosis, the combined structural and functional data from PET and MRI may also improve our understanding of disease pathomechanisms, which may pave the way for developing new integrated imaging biomarkers [11, 12]. Compared to using neuropsychological testing or a single modality, this combined imaging data may assist in better stratifying disease severity and predicting disease progression [13, 14]. As a result, it may be possible to predict response to treatment in AD patients more accurately [6, 15–18]. The natural course of AD is a gradual decline of symptoms, which means that maintenance of the abilities or not producing symptomatic improvement can be regarded as a treatment response. Currently, no biomarker reliably monitors the course of a disease or indicates improvement with treatment [19]. The findings from previous studies are still insufficient. Due to differences in the criteria (such as the definition of response to treatment and imaging cut-off for abnormality) and methods (such as different modalities, imaging and reconstruction protocols, visual versus semiquantitative or quantitative assessment), the results could be heterogeneous. In this research, we investigated the predictive value of treatment response in AD patients using imaging information from FDG brain PET and in combination with other imaging data, including amyloid PET and MRI, using visual and semiquantitative analyses by freely available software.

2 Materials

2.1 Patients

A total of 26 patients, aged ≥60 years, with clinically diagnosed AD, were involved in this study. AD diagnosis was established by the National Institute of Neurological and Communicative Disorders and Stroke and the Alzheimer's Disease and Related Disorders Association (NINCDS-ADRDA) for probable AD [20], TMSE score < 26, and CDR score ≥ 0.5. Subjects were excluded from the study if any psychiatric or neurological illness other than AD was present. Neuropsychological scores at baseline and 1-year follow-up period to assess response to treatment, including Thai Mental State Examination (TMSE) [21], clinical dementia rating (CDR) [22], and Alzheimer's disease assessment scale-cognitive subscale (ADAS-Cog) [23].

2.2 Radiotracers

Following USP chapter 823 guidelines and the CGMP for PET drugs, automated radiosynthesis technology was used at the Cyclotron and Radiochemistry facility of Siriraj Hospital, Mahidol University (Bangkok, Thailand) to produce 18F-FDG and 18F-AV45 [24].

2.2.1 Cyclotron and
Synthesizer

The 20 MeV cyclotron (HM-20S model, Sumitomo Heavy Industries, Ltd., Tokyo, Japan) and F300E synthesizer module (Sumitomo Heavy Industries, Ltd., Tokyo, Japan) were used for radiotracer synthesis.

2.2.2 Reagents and
Equipment for Radiotracer
Synthesis

For details, please *see* Table 1.

**2.3 Standard Brain
Phantom**

The Hoffman 3D brain phantom (Data Spectrum Corporation, Durham, North Carolina) was used for standardizing the gray–white matter contrast of PET imaging data before conducting the clinical study.

**Table 1
Reagents and equipment for ^{18}F-FDG and ^{18}F-AV45 radiotracer syntheses**

^{18}F-FDG	^{18}F-AV45
The Sep-Pak™ plus cartridges (Light Waters Accell™ Plus QMA, PS-2, and Alumina N) from Waters Corporation (MA, USA)	(E)-2-(2-(2-(5-(4-(tert-butoxycarbonyl(methyl) amino) styryl) pyridin-2-yloxy) ethoxy) ethoxy) ethyl-4-methylbenzenesulphonate (AV-105) precursor
The Maxi-Clean SPE 0.5 mL IC-H cartridges from Grace Davison Discovery Science (USA).	0.5 mL of DMSO
Sterile membrane filter (Vented Millex® GS) from Merck Millipore (MA, USA)	0.5 mL of MeCN
Dose calibrator (Capintec, C^{RC-55t})	1 mL of 1 N HCl
Kryptofix 2.2.2 solution (K2.2.2 22 mg, K$_2$CO$_3$ 3.3 mg in 0.7 mL of MeCN and 0.2 mL of H2O)	0.5 mL of 1 N NaOH
0.5 mL of MeCN	Semi-preparative radio-HPLC (Zorbax Eclipse XDB-C18, 5 μm, 9.4 × 250 mm)
Mannose triflate, ultrapure precursor (20 mg/vial)	Phenomenex C18 Security Guard Cartridge, 10 × 10 mm)
KT300 reagent kit (Cryptand 2.2.2 (22 mg/vial), potassium carbonate (K$_2$CO$_3$, 7 mg in water/ acetonitrile, 2:7, 0.9 mL), anhydrous acetonitrile (0.3 and 1.8 mL), sodium hydroxide (0.3 N, 3.0 mL), potassium carbonate solution (16%, 10 mL)) for ^{18}F-FDG synthesis with Sumitomo F300E module from ROTEM Industries Ltd. (Mishor Yamin D.N Arava, Israel)	20 mM NH4OAc pH 4.5 (44:55 v/v)
	0.5% sodium ascorbate
	C18 cartridge (Waters Corp., MA, USA)
	0.22 μm sterile filter (PVDF) (Merck Millipore, MA, USA)
	C18 column (Zorbax Eclipse XDB-C18, 5 μm, 4.6 × 150 mm; Agilent)
^{18}O enriched water (≥99 atom%) from Taiyo Nippon Sanso cooperation (Tokyo, Japan)	Gas chromatography equipped with a FID detector (Agilent Technologies, CA, USA)
Sterile water for injection (Able Medical Company Ltd.)	Limulus amebocyte lysate (LAL) test using the endosafe-PTS (Charles River Laboratory, USA)
0.9% sodium chloride solution (Able Medical Company Ltd.)	
10 mL of 100% ethanol (EtOH)	
10 mL of a 16% potassium carbonate solution	
60 mL of ultrapure water	
0.3 N NaOH	
Sterile product vial	
0.22 μm sterile filter	

2.4 PET/CT Scanner

The Discovery STE PET/CT scanner (GE Healthcare, WI, USA) was used to acquire and reconstruct brain PET/CT images in all patients.

2.5 MRI Scanner

The 3T MR system (Ingenia, Philips Medical System, Best, the Netherlands) was used to acquire brain MRI images in all patients.

3 Methods

The protocol of human subject studies (COA No. Si137/2015) was approved by the Institutional Review Board of the Siriraj Hospital, Mahidol University (Bangkok, Thailand).

3.1 Radiotracer Synthesis

^{18}F-FDG and ^{18}F-AV45 syntheses were controlled by an automated synthesis system and operated in sterile cassettes for the F300E Sumitomo module under a nitrogen atmosphere and 99.9999% purity. A dose calibrator was used to measure the radioactivity.

3.1.1 Production of ^{18}F

Using an HM-20S cyclotron and ^{18}O(p,n)^{18}F reaction on ^{18}O-enriched water as the target material, ^{18}F-fluoride was prepared and transferred to a QMA anion exchange cartridge. A Kryptofix 2.2.2 solution (K2.2.2 22 mg, K_2CO_3 3.3 mg in 0.7 mL of MeCN, and 0.2 mL of H_2O) was used to elute ^{18}F-fluoride. After the K[^{18}F]F-K2.2.2 complex was dried, 0.5 mL of MeCN was added to azeotropically dry under vacuum and nitrogen flow.

3.1.2 Synthesis of ^{18}F-FDG

The CFN synthesis module, mannose triflate, ultrapure precursor (20 mg/vial), and KT300 reagent kit for ^{18}F-FDG synthesis with Sumitomo F300E module and chemical kits from the same company were used to produce ^{18}F-FDG. The reagent kit included Cryptand 2.2.2 (22 mg/vial), potassium carbonate (K_2CO_3, 7 mg in water/acetonitrile, 2:7, 0.9 mL), anhydrous acetonitrile (0.3 and 1.8 mL), sodium hydroxide (0.3 N, 3.0 mL), and potassium carbonate solution (16%, 10 mL). These reagents were used without any further purification.

In preparation for cartridge before use, Light Waters Accell™ Plus QMA cartridges were preconditioned with 10 mL of 100% ethanol (EtOH), followed by 10 mL of a 16% potassium carbonate solution, and then 60 mL of ultrapure water. PS-2, Alumina N, and IC-H cartridges were preconditioned with 10 mL of 100% EtOH and 40 mL of ultrapure water.

The radiosynthesis began with performing a preliminary check of the system's internal parameters, including exhaust test, heating test, pressure test, flow rate of dry nitrogen gas and air test, Sep-Pak cartridge test, and on-off testing of pinch valves for flow and leak detection.

The automated F300E synthesizer receives ^{18}F-fluoride in ^{18}O-enriched water (2.8 mL) via an ^{18}O(p,n)^{18}F reaction. The ^{18}F-fluoride produced by the cyclotron was trapped within a pre-conditioned Sep-Pak Light Waters Accell™ Plus QMA cartridge and subsequently eluted with a K_2CO_3/Cryptand 2.2.2 solution facilitated by a phase transfer catalyst. The azeotropic drying of the crude mixture was achieved using a stream of dry nitrogen under negative pressure to form a dried $[K/Kryptofix] + ^{18}F-$ complex at 100 °C for 6 min before adding anhydrous acetonitrile. The fluori-nation process was initiated by adding mannose triflate precursor (20 mg in 1.8 mL of anhydrous acetonitrile) to this dried crude mixture. The nucleophilic substitution (SN2) reaction was per-formed at 100 °C for 5 min, replacing triflate with fluoride ions to produce acetylated fluorinated glucose. The system then under-went one additional drying at 90 °C for 4 min to remove any excess of acetonitrile impurity. The hydrolysis was accomplished at 100 °C for 5 min by adding 0.3 N NaOH. Subsequently, the crude mixture was cooled down and transferred onto the purification cartridges.

The preconditioned cartridges were used to remove impurities such as Cryptand 2.2.2, free Fluoride ion, K^+, Na^+, CO_3^{2-}, man-nose triflate precursor, and fluorinated mannose triflate precursor. The resulting activity was transferred to the sterile product vial. Furthermore, the final ^{18}F-FDG was eluted with 5 mL of water for injection, passing through the reaction vessel to the refining car-tridges to wash out any remaining radioactivity. This eluate was collected in the sterile product vial with a 0.22 μM sterile mem-brane filter. Subsequently, the individual preparation of the unit dose involved dividing the stock solution and diluting it with a sterile 0.9% sodium chloride solution. The resulting unit dose of ^{18}F-FDG was dispensed into a vented sterile collection vial by passing through a 0.22 μm sterile filter.

3.1.3 Synthesis of ^{18}F-AV45

Following previously published procedures, the (E)-2-(2-(2-(5-(4-(tert-butoxycarbonyl(methyl)amino)styryl)pyridin-2-yloxy)ethoxy)ethyl-4-methylbenzenesulphonate (AV-105) precur-sor was synthesized [25]. The ^{18}F-AV45 radiosynthesis was mod-ified from previously published [26]. A pair of CFN synthesis modules were utilized with radio-HPLC and rotary evaporator modules. To activate the K18F-F-K2.2.2 complex, a solution of AV-105 (5 mg) in DMSO (0.5 mL) and MeCN (0.5 mL) was added. To eliminate a BOC-protecting group, 1 N HCl (1 mL) was added to the mixture and heated to 140 °C for 8.40 min. After 1 min of cooling, the reaction was followed by a PH adjustment with 1 N NaOH (0.5 mL) in the reaction mixture. Semipreparative radio-HPLC was utilized to purify the crude product ^{18}F-AV45 (Zorbax Eclipse XDB-C18, 5 μm, 9.4 × 250 mm guarded with a Phenomenex C18 Security Guard Cartridge, 10 × 10 mm) using a mixture of MeCN and 20 mM NH4OAc pH 4.5 (44:55 v/v) with 0.5% sodium ascorbate as a mobile phase at a flow rate of 4 mL/min with UV detection at 230 nm. The extracted ^{18}F-AV45 (tr = 24 min) was loaded on a C18 cartridge (Waters Corp., MA, USA) after being diluted with H2O (10 mL). EtOH was used to elute the purified ^{18}F-AV45 (2 mL). Then, ^{18}F-AV45 was heated to 95 °C for 3.30 min to minimize the volume of EtOH, and 5 mL of 0.9% sodium chloride solution was added to create the stock solution. Each unit dose was prepared separately by splitting the stock solution, diluting it with a 0.9% sodium chloride solution, and filtering it using a 0.22-μm sterile PVDF filter. An analytical radio-HPLC system using a C18 column (Zorbax Eclipse XDB-C18, 5 μm, 4.6 × 150 mm; Agilent) with ultraviolet detection at 230 nm was used to determine the radiochemical purity of ^{18}F-AV45. The mobile phase was a mixture of MeCN and 20 mM NH4OAc pH 4.5 (55:45 v/v) at a 1 mL/min flow rate [27]. Gas chromatography fitted with an FID detector determined the residual solvents (EtOH, MeCN, and DMSO). Using the endosafe-PTS, the Limulus amebocyte lysate (LAL) test evaluated bacterial endotoxins. Sterility tests were conducted within 2 weeks following production.

Every dose of ^{18}F-FDG and ^{18}F-AV45 produced for this study underwent standard quality control and safety testing before being injected into the patients.

3.2 Standardization of Gray–White Matter Contrast for Quantification of PET Imaging Data

Before starting the clinical trial, the Hoffman 3D brain phantom was used to standardize the gray–white matter contrast for quantification of PET imaging data and assess the image quality generated from the PET/CT scanner and imaging protocols [28]. According to the manufacturer's manual, the ratio of gray-to-white matter radioactivity concentration ratio in this phantom should be 4:1. Using clinical ^{18}F-FDG brain imaging protocol, 0.5 mCi (20 MBq) of the ^{18}F-FDG solution was injected into the phantom during PET scanning with the Discovery STE PET/CT scanner. Following the CT scan, an emission scan was performed continuously for 30 min in list mode. The emission scan was then rearranged as a static image and reconstructed using a 4 iteration, 20 subsets, and no postfiltering 3D-OSEM (VUE point) algorithm. The Hoffman phantom image and digital phantom data were downloaded from the Data Spectrum Corporation website to analyze the image. A 6 mm 3D Gaussian filter FWHM was then used to calculate the gray–white contrast. Using the processing program PMOD 3.7 (PMOD Technologies, Zurich, Switzerland), the regions of interest (ROIs) were created on the digital Hoffman phantom according to the JSNM ROI templates. Digital and PET phantom images were successfully coregistered (Fig. 1), and the percentage contrast was computed as:

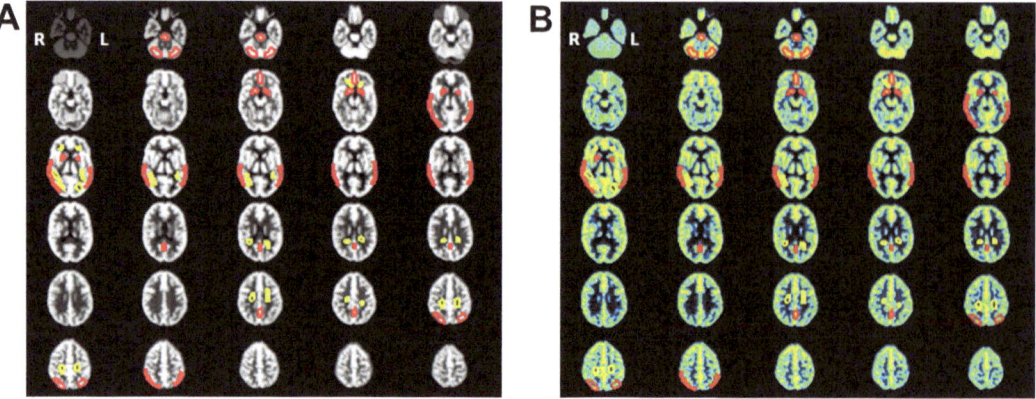

Fig. 1 Standardization of gray–white matter contrast for quantification of PET imaging data using Hoffman brain phantom. (First published by Thientunyakit et al. [14] by Springer Nature with reprint permission)

$$\%\text{Contrast} = \frac{(\text{GMp}/\text{WMp} - 1)}{(\text{GMd}/\text{WMd} - 1)} \times 100$$

The gray and white matter ROI activities of the phantom PET image were represented by the GMp and WMp, respectively, and the GMd and WMd represented the ROI activities of the digital phantom. The GMp and WMp from phantom PET image were 12.161 and 4.474, respectively. The manufacturer technical data of the GMd and WMd ratio was 4:1. The above percentage contrast formula's substitution produced a 57.276% contrast.

3.3 Clinical Imaging Protocols

All ^{18}F-FDG PET/CT, ^{18}F-AV45 PET/CT, and 3D T1-weighted MRI studies of the brain at baseline were performed within 6 weeks after the neurocognitive tests. The PET/CT imaging protocols are summarized in Table 2.

3.3.1 ^{18}F-FDG Brain PET/CT

Dynamic PET/CT imaging of the brain was performed with Discovery STE PET/CT scanner at 30 min after intravenous administration of the ^{18}F-FDG radiopharmaceutical (radioactivity dose range 4.5–5.5 mCi or approximately 166.5–203.5 MBq) for 30 min duration, following the scanner-specific protocol from Alzheimer's Disease Neuroimaging Initiative (ADNI) [29].

3.3.2 ^{18}F-AV45 Brain PET/CT

Dynamic PET/CT imaging of the brain was performed with the same PET/CT scanner at 50 min after intravenous administration of ^{18}F-AV45 (radioactivity dose range 8–10 mCi) for 20 min duration, following the scanner-specific protocol from Alzheimer's Disease Neuroimaging Initiative (ADNI) [29].

Table 2
Summary of scanner-specific PET/CT imaging protocols [29]

Radiotracers	18F-FDG	18F-AV45
Dose (9 mCi)	4.5–5.5	8–10
Uptake period (min)	30	50
Mode	Dynamic 3D mode, 5 min/frame	
Total frames	6	4
Scan period (min)	30	20
Reconstruction method	3D iterative reconstruction (VUE point)	
Matrix size	128*128	
Voxel size (mm)	2	
CT slide thickness (mm)	3.27	
Attenuation correction	CT-base attenuation correction	
Reconstruction	Ordered subsets expectation-maximization (OSEM) 4 iterations, 20 subsets	

3.3.3   3D T1-Weighted Brain MRI

MRI was performed on a 3T MR system with a 32-channel head coil. The imaging protocol included a 3D high-resolution T1-weighted FFE covering the whole brain (field of view [FOV] $230 \times 230 \times 172$ mm^3, matrix size 352×352, voxel size $0.72 \times 0.72 \times 0.65$ mm^3, echo time TE/repetition time TR 4.8/9.8 ms; flip angle 8°, scan time 6 min).

3.4   Brain Imaging Data Analysis

3.4.1   Visual Analysis

The visual analysis of the brain PET image was interpreted by two experienced nuclear medicine physicians in consensus. The visual analysis of brain MRI images was performed by one experienced neuroradiologist.

18F-FDG Brain PET Images

The patterns of brain glucose metabolism were assessed using the original images [30] and three-dimensional stereotactic surface projections (3D-SSP) Z-score map images from NEUROSTAT software [31] and described as:

– Normal (score 0) = No alteration of brain glucose metabolism.
– Early AD (score 1) = Hypometabolism in posterior cingulate cortex.
– Mild AD (score 2) = Early AD findings plus mild temporal–parietal hypometabolism.
– Moderate AD (score 3) = Early AD findings plus severe temporal–parietal hypometabolism with normal frontal metabolism.
– Severe AD (score 4) = Moderate AD findings plus frontal hypometabolism.

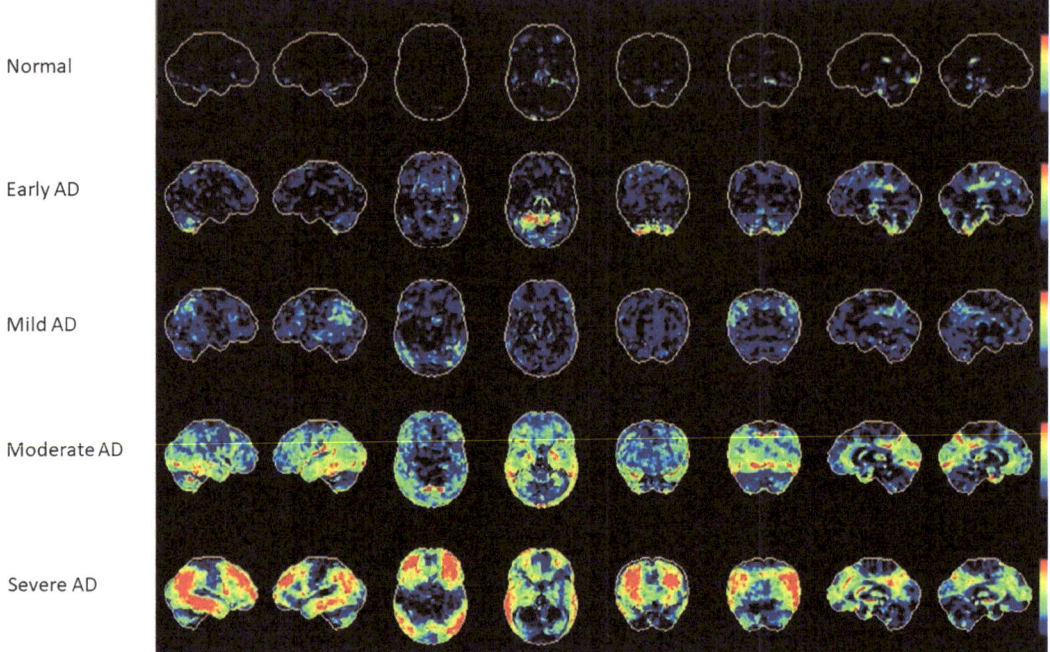

Fig. 2 ^{18}F-FDG 3D-SSP Z-score map patterns of normal and Alzheimer's disease (AD) patients with different severity based on areas with glucose hypometabolism

The normal pattern (score 0) was then interpreted as negative FDG PET, and abnormal scans (score 1–4) were interpreted as positive FDG PET. The AD patterns from 3D-SSP Z-score map images are identified in Fig. 2.

^{18}F-AV45 Brain PET Images   The ^{18}F-AV45 images are interpreted following previous recommendations [32–35]. A negative amyloid PET image was considered when the radiotracer uptake in the cortical white matter was more than in the gray matter, resulting in a clear gray and white matter contrast. Positive amyloid PET image was considered when there was significant radiotracer uptake in the cortical gray matter, which resulted in loss of gray and white matter contrast by either of these findings: (A) the intensity in gray matter is equal to that in the adjacent white matter, involved at least two regions, where each region involved more than one gyrus, or (B) the uptake intensity in gray matter is more than in the adjacent white matter at least one region [35, 36]. Semiquantitative 3D-SSP Z-score map images were also used to aid visual assessment from the original image and grading into five stages, with stage II–IV considered positive brain amyloid deposition [37] (Fig. 3a, b).

– Stage 0 = No abnormal uptake.
– Stage I = Uptake in the basal part of the temporal lobe, the anterior cingulate gyrus, and the parietal operculum.

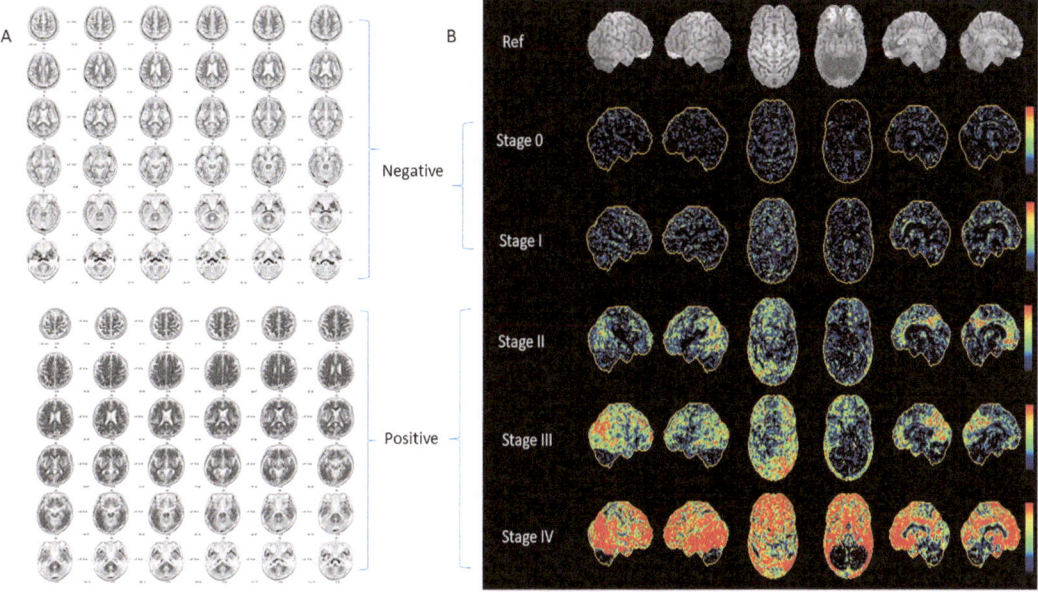

Fig. 3 Examples of ^{18}F-AV45 positron emission tomography (PET)/CT scan to demonstrate amyloid negative vs positive images from original axial images (**a**) and 3D-SSP Z-score map images (**b**). Abnormal amyloid PET was considered in the regions with a Z-score level of 2 or more, corresponding to green color and above on the color scale bar. Stages II or more are considered positive amyloid PET scans. (Modified from Thientunyakit et al. [37] by Malaysian Medical Association with reprint permission)

– Stage II = Uptake in vast parts of the temporal, frontal, and parietal associative cortex.

– Stage III = Uptake in primary sensory-motor cortices and anterior medial temporal lobe.

– Stage IV = Uptake in the posterior medial temporal lobe and the striatum.

Brain MRI Images   T1-weighted MRI images were assessed for Scheltens' visual rating scale or medial temporal lobe atrophy (MTA) score [38] using the width of the choroid fissure and temporal horn and the height of hippocampus as a 5-point rating scale as follows (Fig. 4):

– Score 0 = No atrophy.

– Score 1 = Only widening of choroid fissure.

– Score 2 = Also widening of the temporal horn of the lateral ventricle.

– Score 3 = Moderate loss of hippocampal volume (decrease in height).

– Score 4 = Severe volume loss of hippocampus.

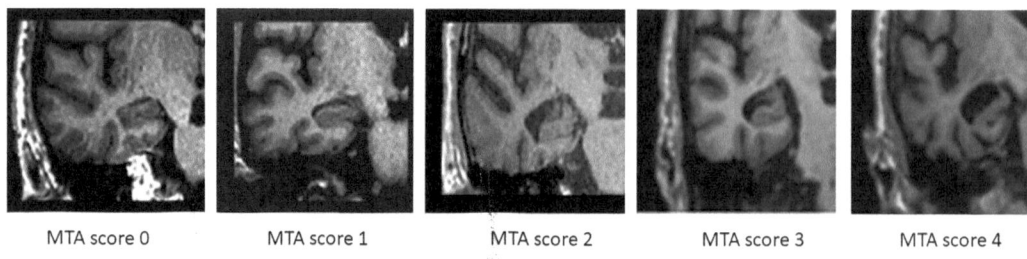

MTA score 0   MTA score 1   MTA score 2   MTA score 3   MTA score 4

Fig. 4 Scheltens' visual rating scale or medial temporal lobe atrophy (MTA) score based on the width of the choroid fissure and temporal horn and the height of the hippocampus [38]

The cut-off values for positive hippocampal atrophy concerning the average right and left MTA scores for the age range [39] were as follows:

– Age < 65 years cut-off values ≥1.0.

– Age 65–74 years cut-off values ≥1.5.

– Age ≥ 75 years cut-off values ≥2.0.

Integrated Visual Score Index

The visual assessment of PET/CT and MRI images is combined to grade the severity of neurodegenerative brain changes as follows:

– Score 0 = negative amyloid, FDG, and MRI (A − F − M−).

– Score 1 = positive amyloid, negative FDG, and MRI (A + F − M−).

– Score 2 = positive amyloid and FDG, negative MRI (A + F + M−).

– Score 3 = positive amyloid, FDG, and MRI (A + F + M+).

3.4.2 Quantitative Analysis

Each patient's brain PET and MRI images acquired with ^{18}F-FDG and ^{18}F-AV45 were quantitatively analyzed using an automated program, FreeSurfer Imaging Analysis Suite version 6.0.0. To obtain respective cortical brain volume from the MRI image and standardized uptake volume (SUV) from PET images, the patient's MRI image was coregistered with the MRI template for further automated segmentation and parcellation of the cortical gray and white matter into 147 volumes of interest (VOIs) following Free-Surfer's pipeline [40] (Fig. 5).

Semiquantitative Analysis of ^{18}F-FDG and ^{18}F-AV45 PET Data

NEUROSTAT Workflow

A 3D-SSP Z-score map image for each patient's ^{18}F-FDG and ^{18}F-AV45 was generated with the Neurological Statistical Image Analysis Program (NEUROSTAT/3D-SSP program, University of Utah, SLC, USA). The ImageJ 1.51 s software (National Institutes of Health, USA, available at http://imagej.nih.gov/ij) was used to convert anonymized summed DICOM image files of each PET scan into a single analyze format file. NEUROSTAT software was

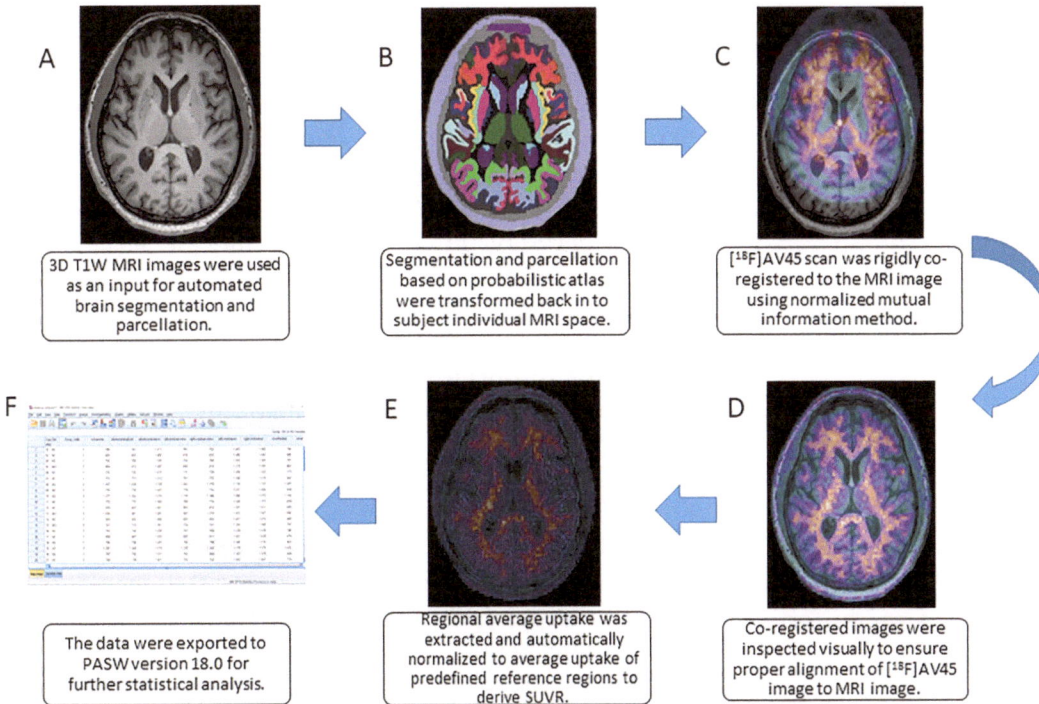

Fig. 5 The diagram of quantitative assessment for 18F-AV45, 18F-FDG, and T1-weighted (T1W) MRI images using FreeSurfer imaging analysis suite version 6.0.0 (**a**) T1W image, (**b**) automated segmentation and parcellation of brain cortex using the standard brain template, (**c**) PET image coregistration with patient's MRI, (**d**) coregistered PET and MRI images to check for the suitable alignment, (**e**) automated segmentation of PET and MRI images, and (**f**) calculation of regional SUVR. (First published by Thientunyakit et al. [14] by Springer Nature with reprint permission)

utilized to process the converted files by transforming reconstructed images to the stereotactic coordinate system and coregistering 18F-FDG and 18F-AV45 PET images [31, 37, 41]. An FDG template performed anatomical standardization (AS) on the 18F-FDG image. Then, the 18F-AV45 and 18F-FDG images were coregistered. AS was done using the transformation parameters calculated for AS of the corresponding 18F-FDG image. 18F-AV45 images could be transformed into the stereotactic space of the standard brain by applying the library files for the 18F-FDG image generated by the NEUROSTAT algorithms to the 18F-AV45 image. Before further analysis, each image obtained by the NEUROSTAT software was reviewed for quality control for the alignment and coregistration of both PET scans. Like 18F-FDG PET, the regional activities 18F-AV45 were obtained from cortical gray matter to the template's surface using the 3D-SSP approach. Using NEUROSTAT scaling techniques, the SUVR was computed from cortical activity in specific brain regions normalized by the average activity of the cerebellar cortex.

FreeSurfer Workflow

SUVR and volumetric analysis were obtained regionally using FreeSurfer software. Using the PETSurfer workflow in FreeSurfer, the ^{18}F-FDG and ^{18}F-AV45 PET images were coregistered to high-resolution segmented brain MR images [42, 43]. The coregistered PET/MRI scans were visually examined to verify correct alignment. Next, for each gyral and lobar cortical region and subcortical structure, the average SUV of ^{18}F-FDG was retrieved. The extracted average SUV was normalized using the average SUV of the cerebellar cortex to produce the ^{18}F-FDG SUVR [44, 45]. The average SUV of each gyral and lobar cortical region and subcortical structure was normalized to the average SUV of the cerebellar cortex to obtain the SUVR of ^{18}F-AV45. The SUVR of each region was computed using the following formula.

$$\text{SUVR (interested VOI)} = \text{SUV (interested VOI)}/\text{SUV (reference VOI)}$$

Volumetric Analysis of MRI Data

Using FreeSurfer image analysis suite version 6.0.0, the 3D high-resolution T1-weighted FEE images were processed using automated brain segmentation, cortical parcellation, and skull striping to get the normalized volume (NVol) for each cortical region, with technical details described in previous publications [46]. Each cerebral hemisphere's cortical parcellation into 34 gyral subregions was based on the Desikan–Killiany brain atlas [40]. Based on the criteria outlined in the brain atlas, these subregions were summed to form larger cortical lobar regions: frontal, parietal, temporal, occipital, insula, and the whole cingulate (anterior and posterior regions). Thirty-seven subcortical regions, including the caudate, putamen, thalamus, hippocampus, amygdala, n. accumbens, whole cingulate gurus (both anterior and posterior), and other regions, were also identified by subsegmenting subcortical structures. The MRI voxel volume was multiplied by the number of voxels in each segmented section to determine the volume of that region. Each region's volume was divided by the estimated total intracranial volume (eTIV) to calculate the eTIV-corrected volume. Normalized volume was obtained by normalizing the eTIV-corrected volume of each region using the average eTIV-corrected volume of each region from the age-matched control group.

The NVol of each cortical region was computed using the following formula:

$$\text{Volume for MRI (mapping to template)} = \text{Normalized volume (nVol)}$$

3.4.3 The Amyloid/FDG/ nVol Index

As in our previous work, the quantitative data from ^{18}F-AV45, ^{18}F-FDG, and MRI were integrated into a single value: amyloid/FDG/nVol index [14]. This research's specific regions of interest were frontal, parietal, temporal, occipital, posterior cingulate, precuneus, hippocampus, and composite regions [47] (Table 3). A composite VOI was calculated by combining 38-gyral VOIs from bilateral major cortical regions (frontal, temporal, parietal, precuneus, and whole cingulate) [29] and normalized by the corresponding average composite volume of the healthy age-matched control group (composite nVol). SUVR of the composite VOI for ^{18}F-FDG and ^{18}F-AV45 was calculated similarly to that of the abovementioned VOIs.

We calculated AV45/FDG/NVol indices and generated parametric and z-score images based on regional AV45/FDG/NVol indices. ^{18}F-AV45 SUVR for each region was divided by ^{18}F-FDG SUVR for the same region to derive the AV45/FDG index. AV45/FDG/NVol index for a region was calculated by dividing the AV45/FDG index by NVol for the same region.

3.5 Outcome Measurement

Response to treatment was assessed using the changes in neuropsychological assessment scores, including TMSE, ADAS-cog, and CDR, between baseline and a 1-year follow-up period. Improved or stable scores of all neuropsychological tests defined treatment responder (TR). Nonresponder (NR) was defined by a worsening score in at least one test (ADAS-cog increase of 4 scores or more, TMSE decrease of 1 score or more, or CDR increase of 0.5 score or more).

3.6 Statistical Analyses

Fisher's exact test for continuous data and χ^2 for categorical data were used to characterize study groups' demographic and neurocognitive performance data. The association between baseline imaging results and response to treatment was assessed using relative risk. The receiver operating characteristic (ROC) curve was used for diagnostic performance analysis and to identify each imaging result's cut-off value. Multivariable analysis was performed using generalized linear regression with a log link and binomial

Table 3
The cortical gray matter regions and subregions included as representative of composite region [29, 47]

Regions	Subregions
Frontal	Caudate middle frontal, lateral orbital frontal, medial orbital frontal, pars opercularis, pars orbitalis, pars triangularis, rostal middle frontal, superior frontal, and frontal pole
Anterior/posterior cingulate	Caudate anterior cingulate, rostal anterior cingulate, isthmus cingulate, and posterior cingulate
Lateral parietal	Superior parietal, inferior parietal, supramarginal, and precuneus
Lateral temporal	Superior temporal and middle temporal

distribution. Statistical significance was defined as $P < 0.05$. Statistical analyses were accomplished using PASW Statistics for Windows, Version 18.0 (SPSS Inc., Chicago, IL, USA).

---

4   Notes

4.1   Major Problems or Faults with the Techniques Used and How to Identify and Overcome

4.1.1   Patients

This study's small number of AD patients led to insufficient statistical power. The estimated sample size should be 148 patients based on the earlier research by Um et al. [15] and Adler et al. [6], which had the expected sensitivity of 80%, specificity of 70%, positive likelihood ratio of 2, and treatment response rate of 0.59 and 0.4, respectively. However, this study aimed to evaluate data from multimodal imaging beyond ^{18}F-FDG PET. Future studies with a larger population can benefit from the preliminary findings of this pilot study.

Another drawback of this study is the lack of a gold standard for confirming AD pathology. Brain autopsy and CSF data are missing, and only clinical diagnosis of AD by NINCDS-ADRDA criteria was available. Only during the postmortem phase can a brain autopsy be performed, and many patients and their families still view lumbar puncture for CSF study as an invasive procedure. It is well established that clinical criteria and neuropsychological testing have limited accuracy in diagnosis. This limitation may account for the negative visual ^{18}F-florbetapir PET results in 4 out of 26 patients with suspected AD (TR 2, NR 2). A low level of amyloid deposition may be under the detectable threshold by PET signals; however, it may not explain the dementia symptoms due to AD.

Confounding variables could impact neurocognitive function, and consequently, the response to treatment includes vascular risk factors, other psychological illnesses (such as depression), and partial treatment with other medications or interventions. To identify patients with confounding factors, we thoroughly reviewed each patient's medical record and incorporated this information into the questionnaires. Patients with poorly controlled diabetes may express poor image quality of ^{18}F-FDG PET image. Individuals with poorly managed diabetes may exhibit low-quality ^{18}F-FDG PET images. Except for acceptable controlled vascular risk factors in about 34% of patients (diabetes, hypertension, and dyslipidemia), none of the mentioned confounding factors were evident in our patients. No patients demonstrated significant vascular comorbidity in the correlative MRI image. We also confirmed that no patients received dementia medications or interventions other than AchEIs and NMDA antagonists.

4.1.2   Image Acquisition and Reconstruction

Significant patient head motion was detected during the 30-min ^{18}F-FDG PET/CT scan in one patient. The detection was made by continuously monitoring the patient and sinogram during acquisition and checking the reconstructed PET images immediately after the scan. We used list-mode acquisition, which provides this patient's solution by excluding image frames affected by head motion and summing up the remaining frames with good quality for further analysis.

4.1.3 Image Analysis

NEUROSTAT Software

The NEUROSTAT software requires files in the header (*.hdr) and image (*.img) formats. Therefore, we need another software for converting the original DICOM image file to those formats, which requires some technical skills and quality control of the converted file to ensure the results' validity. However, the new version of NEUROSTAT software can directly analyze the DICOM image for FDG PET and some particular studies but currently does not include amyloid PET.

There was misregistration between ^{18}F-FDG and ^{18}F-AV45 PET images in one patient with deficient uptake of ^{18}F-AV45 in the brain and high uptake in the scalp or parotid glands, which resulted in an artifact in the ^{18}F-AV45 Z-score map image. The misregistration was detected during the QC process and corrected by modifying some parameters in the analysis command.

FreeSurfer Software

The automatic workflow failed during the segmentation of the MRI images in two cases by mis-differentiating between cerebrospinal fluid (CSF) and gray matter, and between gray and white matter. This issue was detected during the QC process and corrected by point-by-point manual editing. However, there was one case with severe white matter pathology causing a similar signal between gray and white matter, resulting in globally incorrect segmentation, which could not be corrected manually. We needed to exclude this case from the study.

One concern about using FreeSurfer software, which requires the patient's MRI and PET images, is the partial volume (PVC) or spillover effect due to the difference in image resolution between the two modalities. PVC may degrade the quantitative accuracy of PET images because a voxel's intensity reflects its tracer concentration and surrounding area. However, we used the PVC tools provided by FreeSurfer to correct for spillover effects resulting from PET images' restricted spatial resolution.

4.2 Variations of the Technique

We had to remove the frames with significant head motion affecting PET image quality in one patient. However, the remaining image frames displayed sufficient count statistics and optimal image quality for further analysis and interpretation. Furthermore, this study divided the regional uptake counts by the uptake in the reference region to apply normalized SUVR to minimize the effect of low-count statistics.

4.3 Sensitivity of the Methods

Software for quantitative and semiquantitative analysis is widely accessible to improve the diagnostic efficacy of visual interpretation. Semiquantitative analysis of FDG PET images reported a sensitivity as high as 95–97% and a specificity of 100% in discriminating patients with probable AD from normal subjects [48]. However, discordant results between visual and quantitative assessments occasionally represent a significant issue. The highly heterogeneous FDG PET sensitivity and specificity results in previous studies might be explained by different criteria and differences in methodological approaches. Furthermore, overlapping FDG PET abnormality among neurodegenerative diseases and mixed pathologies also affects its diagnostic performance [49, 50]. For amyloid PET, our previous work demonstrated a higher agreement of results from semiquantitative analysis than visual assessment compared to clinical diagnosis ($R^2 = 0.793$ and 0.845, respectively). The stages of amyloid deposition using Z-score maps are also consistent with neurocognitive status across subjects, although not ideally (stage 0 in 95% of HC, stage I in 3.3% of AD, and stage II–IV in 80% of AD) [37]. These findings again raised the question of the accuracy of clinical diagnosis.

5   Conclusion

The predictive value of treatment response to medications in AD patients using multimodal PET and MRI imaging findings to evaluate cerebral glucose hypometabolism alongside amyloid-beta deposition and morphological change may be a promising approach to overcome previous limitations from using clinical factors, neuropsychological score, or single imaging modality. With the known low response to medical treatment, knowing disease severity from molecular and anatomical imaging is crucial in helping select suitable patients. While additional semiquantitative and quantitative analyses are well accepted for their higher diagnostic performance than visual analysis alone, they require technical expertise and intensive quality control checks. Integrating both visual and quantitative techniques also possibly presents discordant results. Therefore, we emphasize that researchers should be aware of these technical issues and set a clear definition and cut-off for image interpretation.

Acknowledgments

The authors thank Professor Opa Vajragupta and Associate Professor Supatporn Tepmongkol for their helpful comments, Angkana Jongsawadipatana, Phattaranan Phawaphutanon and Dollaporn Polyeam for data collection, and the staff at the Siriraj Cyclotron

Center and Imaging Center, Department of Radiology and Department of Preventive and Social Medicine. This project is cofunded by the National Research Council of Thailand (NRCT) through the Health System Research Institute (HSRI), the Faculty of Medicine Siriraj Hospital, Mahidol University, and the International Atomic Energy Agency (IAEA).

Disclosures None of the authors has any conflicts of interest in this study.

References

1. Morris JC (1996) Classification of dementia and Alzheimer's disease. Acta Neurol Scand 94(S165):41–50

2. World Health Organization (2021) Global status report on the public health response to dementia. Available at https://www.who.int/publications-detail-redirect/9789240033245

3. Birks J (2006) Cholinesterase inhibitors for Alzheimer's disease. Cochrane Database Syst Rev 2006(1):CD005593

4. Lanctôt KL, Herrmann N, Yau KK, Khan LR, Liu BA, LouLou MM et al (2003) Efficacy and safety of cholinesterase inhibitors in Alzheimer's disease: a meta-analysis. CMAJ 169(6): 557–564

5. Farlow M, Anand R, Messina J Jr, Hartman R, Veach J (2000) A 52-week study of the efficacy of rivastigmine in patients with mild to moderately severe Alzheimer's disease. Eur Neurol 44(4):236–241

6. Adler G, Brassen S, Chwalek K, Dieter B, Teufel M (2004) Prediction of treatment response to rivastigmine in Alzheimer's dementia. J Neurol Neurosurg Psychiatry 75(2):292–294

7. Pauwels E, Sturm E, Bombardieri E, Cleton F, Stokkel M (2000) Positron-emission tomography with [18 F] fluorodeoxyglucose: Part I. Biochemical uptake mechanism and its implication for clinical studies. J Cancer Res Clin Oncol 126:549–559

8. Mosconi L (2005) Brain glucose metabolism in the early and specific diagnosis of Alzheimer's disease: FDG-PET studies in MCI and AD. Eur J Nucl Med Mol Imaging 32(4): 486–510

9. Chételat G, Arbizu J, Barthel H, Garibotto V, Law I, Morbelli S et al (2020) Amyloid-PET and 18F-FDG-PET in the diagnostic investigation of Alzheimer's disease and other dementias. Lancet Neurol 19(11):951–962

10. Jack CR, Knopman DS, Jagust WJ, Shaw LM, Aisen PS, Weiner MW et al (2010) Hypothetical model of dynamic biomarkers of the Alzheimer's pathological cascade. Lancet Neurol 9(1):119–128

11. Drzezga A, Barthel H, Minoshima S, Sabri O (2014) Potential clinical applications of PET/MR imaging in neurodegenerative diseases. J Nucl Med 55(Supplement 2):47S–55S

12. Barthel H, Schroeter ML, Hoffmann K-T, Sabri O (eds) (2015). PET/MR in dementia and other neurodegenerative diseases. Semin Nucl Med 45(3):224–233

13. Xu L, Wu X, Li R, Chen K, Long Z, Zhang J et al (2016) Prediction of progressive mild cognitive impairment by multi-modal neuroimaging biomarkers. J Alzheimers Dis 51(4): 1045–1056

14. Thientunyakit T, Sethanandha C, Muangpaisan W, Chawalparit O, Arunrungvichian K, Siriprapa T et al (2020) Relationships between amyloid levels, glucose metabolism, morphologic changes in the brain and clinical status of patients with Alzheimer's disease. Ann Nucl Med 34(5):337–348

15. Um YH, Kim T-W, Jeong J-H, Seo H-J, Han J-H, Hong S-C et al (2017) Prediction of treatment response to donepezil using automated hippocampal subfields volumes segmentation in patients with mild Alzheimer's disease. Psychiatry Investig 14(5):698–702

16. Ohnishi T, Sakiyama Y, Okuri Y, Kimura Y, Sugiyama N, Saito T et al (2014) The prediction of response to galantamine treatment in patients with mild to moderate Alzheimer's disease. Curr Alzheimer Res 11(2):110–118

17. Shimada A, Hashimoto H, Kawabe J, Higashiyama S, Kai T, Kataoka K et al (2011) Evaluation of therapeutic response to donepezil by positron emission tomography. Osaka City Med J 57(1):11–19

18. Tateno A, Sakayori T, Kawashima Y, Higuchi M, Suhara T, Mizumura S et al (2015) Comparison of imaging biomarkers for Alzheimer's disease: amyloid imaging with [18F] florbetapir positron emission tomography and magnetic resonance imaging voxel-based analysis for entorhinal cortex atrophy. Int J Geriatr Psychiatry 30(5):505–513

19. Burns A, Yeates A, Akintade L, Del Valle M, Zhang RY, Schwam EM et al (2008) Defining treatment response to donepezil in Alzheimer's disease: responder analysis of patient-level data from randomized, placebo-controlled studies. Drugs Aging 25(8):707–714

20. The National Institute on Aging, and Reagan Institute Working Group on Diagnostic Criteria for the Neuropathological Assessment of Alzheimer's Disease (1997) Consensus recommendations for the postmortem diagnosis of Alzheimer's disease. Neurobiol Aging 18(4 Suppl):S1–S2

21. Committee TTBF (1993) Thai mental state examination (TMSE). Siriraj Hosp Gaz 45:359–374

22. Morris JC (1993) The Clinical Dementia Rating (CDR): current version and scoring rules. Neurology 43(11):2412–2414

23. Senanarong V, Harnphadungkit K, Poungvarin N, Vannasaeng S, Chongwisal S, Chakorn T et al (2013) The dementia and disability project in Thai elderly: rational, design, methodology and early results. BMC Neurol 13(1):1–11

24. Schwarz SW, Dick D, Van Brocklin HF, Hoffman JM (2014) Regulatory requirements for PET drug production. J Nucl Med 55(7):1132–1137

25. Zhang W, Oya S, Kung M-P, Hou C, Maier DL, Kung HF (2005) F-18 Polyethyleneglycol stilbenes as PET imaging agents targeting Aβ aggregates in the brain. Nucl Med Biol 32(8):799–809

26. Yao C-H, Lin K-J, Weng C-C, Hsiao T, Ting Y-S, Yen T-C et al (2010) GMP-compliant automated synthesis of [18F] AV-45 (Florbetapir F 18) for imaging β-amyloid plaques in human brain. Appl Radiat Isotopes 68(12):2293–2297

27. Liu Y, Zhu L, Plössl K, Choi SR, Qiao H, Sun X et al (2010) Optimization of automated radiosynthesis of [18F] AV-45: a new PET imaging agent for Alzheimer's disease. Nucl Med Biol 37(8):917–925

28. Ikari Y, Akamatsu G, Nishio T, Ishii K, Ito K, Iwatsubo T et al (2016) Phantom criteria for qualification of brain FDG and amyloid PET

across different cameras. EJNMMI Phys 3(1):1–18

29. Jagust WJ, Landau SM, Koeppe RA, Reiman EM, Chen K, Mathis CA et al (2015) The Alzheimer's disease neuroimaging initiative 2 PET core: 2015. Alzheimers Dement 11(7):757–771

30. Brown RK, Bohnen NI, Wong KK, Minoshima S, Frey KA (2014) Brain PET in suspected dementia: patterns of altered FDG metabolism. Radiographics 34(3):684–701

31. Minoshima S, Frey KA, Koeppe RA, Foster NL, Kuhl DE (1995) A diagnostic approach in Alzheimer's disease using three-dimensional stereotactic surface projections of fluorine-18-FDG PET. J Nucl Med 36(7):1238–1248

32. Johnson KA, Sperling RA, Gidicsin CM, Carmasin JS, Maye JE, Coleman RE et al (2013) Florbetapir (F18-AV-45) PET to assess amyloid burden in Alzheimer's disease dementia, mild cognitive impairment, and normal aging. Alzhiemers Dement 9(5):S72–S83

33. Wong DF, Rosenberg PB, Zhou Y, Kumar A, Raymont V, Ravert HT et al (2010) In vivo imaging of amyloid deposition in Alzheimer disease using the radioligand 18F-AV-45 (florbetapir F 18). J Nucl Med 51(6):913–920

34. Lilly E (2012) Amyvid (florbetapir F 18 INJECTION): highlights of prescribing information. Eli Lilly, Indianapolis

35. Clark CM, Schneider JA, Bedell BJ, Beach TG, Bilker WB, Mintun MA et al (2011) Use of florbetapir-PET for imaging β-amyloid pathology. JAMA 305(3):275–283

36. Camus V, Payoux P, Barré L, Desgranges B, Voisin T, Tauber C et al (2012) Using PET with 18 F-AV-45 (florbetapir) to quantify brain amyloid load in a clinical environment. Eur J Nucl Med Mol Imaging 39(4):621–631

37. Thientunyakit T, Sethanandha C, Muangpaisan W, Minoshima S (2021) 3D-SSP analysis for amyloid brain PET imaging using 18F-florbetapir in patients with Alzheimer's dementia and mild cognitive impairment. Med J Malaysia 76(4):493–501

38. Scheltens P, Leys D, Barkhof F, Huglo D, Weinstein H, Vermersch P et al (1992) Atrophy of medial temporal lobes on MRI in "probable" Alzheimer's disease and normal ageing: diagnostic value and neuropsychological correlates. J Neurol Neurosurg Psychiatry 55(10):967–972

39. Claus JJ, Staekenborg SS, Holl DC, Roorda JJ, Schuur J, Koster P et al (2017) Practical use of visual MTA cut-off scores in Alzheimer's disease: validation in a large memory clinic population. Eur Radiol 27(8):3147–3155

40. Desikan RS, Ségonne F, Fischl B, Quinn BT, Dickerson BC, Blacker D et al (2006) An automated labeling system for subdividing the human cerebral cortex on MRI scans into gyral based regions of interest. NeuroImage 31(3):968–980

41. Minoshima S, Koeppe RA, Frey KA, Kuhl DE (1994) Anatomic standardization: linear scaling and nonlinear warping of functional brain images. J Nucl Med 35(9):1528–1537

42. Greve DN, Salat DH, Bowen SL, Izquierdo-Garcia D, Schultz AP, Catana C et al (2016) Different partial volume correction methods lead to different conclusions: an (18)F-FDG-PET study of aging. NeuroImage 132:334–343

43. Greve DN, Svarer C, Fisher PM, Feng L, Hansen AE, Baare W et al (2014) Cortical surface-based analysis reduces bias and variance in kinetic modeling of brain PET data. NeuroImage 92:225–236

44. Minoshima S, Frey KA, Foster NL, Kuhl DE (1995) Preserved pontine glucose metabolism in Alzheimer disease: a reference region for functional brain image (PET) analysis. J Comput Assist Tomogr 19(4):541–547

45. Ishii K, Willoch F, Minoshima S, Drzezga A, Ficaro EP, Cross DJ et al (2001) Statistical brain mapping of 18F-FDG PET in Alzheimer's disease: validation of anatomic standardization for atrophied brains. J Nucl Med 42(4):548–557

46. Fischl B, Salat DH, Busa E, Albert M, Dieterich M, Haselgrove C et al (2002) Whole brain segmentation: automated labeling of neuroanatomical structures in the human brain. Neuron 33(3):341–355

47. Landau SM, Breault C, Joshi AD, Pontecorvo M, Mathis CA, Jagust WJ et al (2013) Amyloid-β imaging with Pittsburgh compound B and florbetapir: comparing radiotracers and quantification methods. J Nucl Med 54(1):70–77

48. Burdette JH, Minoshima S, Vander Borght T, Tran DD, Kuhl DE (1996) Alzheimer disease: improved visual interpretation of PET images by using three-dimensional stereotaxic surface projections. Radiology 198(3):837–843

49. Minoshima S, Mosci K, Cross D, Thientunyakit T (2021) Brain [F-18]FDG PET for clinical dementia workup: differential diagnosis of Alzheimer's disease and other types of dementing disorders. Semin Nucl Med 51(3):230–240

50. Minoshima S, Cross D, Thientunyakit T, Foster NL, Drzezga A (2022) 18F-FDG PET imaging in neurodegenerative dementing disorders: insights into subtype classification, emerging disease categories, and mixed dementia with copathologies. J Nucl Med 63 (Suppl 1):2S–12S

Chapter 7

Opioid Receptor and Advanced Nuclear Imaging in Epilepsy

Daichi Sone and Matthias Koepp

Abstract

In epilepsy, nuclear neuroimaging is an established modality in clinical practice to identify interictal hypometabolic or ictal hyperperfusion areas around the epileptogenic foci. Specific molecular imaging has also been used for scientific and clinical purposes beyond detecting changes in glucose metabolism and locating seizure foci. In this chapter, we first review past studies of the epileptic brain that have used advanced nuclear imaging with various neurotransmitter ligands, including ligands for γ-aminobutyric acid (GABA) and glutamate receptors and monoamines, as well as nuclear imaging tools to evaluate neuroinflammation and blood–brain barrier (BBB) integrity; then, we focus on a methodology for opioid receptor imaging in epilepsy.

Key words Epilepsy, Positron emission tomography, Nuclear imaging, Opioid receptors, Molecular neuroimaging

1 Introduction

Epilepsy is a common brain disorder characterized by recurrent seizures due to abnormal and excessive neural discharges, affecting over 40 million people all over the world [1, 2]. The burden of epilepsy ranges from the risk of accidents and injuries due to seizures to the various comorbidities and psychosocial issues associated with the disease. Thus, in treating patients with epilepsy, it is important to address comorbidities in addition to ensuring appropriate seizure control.

Nuclear imaging is an essential modality in the current clinical approach to epilepsy. While epileptic seizures can be well-controlled by appropriate antiseizure medications in about 70% of cases [3], the remaining cases develop pharmacoresistant epilepsy. In these cases, further investigations and interventions are required, including neurosurgical resection of the focal lesion [4]. At this stage, nuclear neuroimaging is a clinically established examination for locating the epileptic focus. ^{18}F-fluorodeoxyglucose positron emission tomography (^{18}F-FDG-PET) is useful for measuring cerebral

Daichi Sone (ed.), *Molecular Imaging for Brain Diseases*, Neuromethods, vol. 222, https://doi.org/10.1007/978-1-0716-4494-2_7,
© The Author(s), under exclusive license to Springer Science+Business Media, LLC, part of Springer Nature 2025

glucose metabolism related to neuronal and synaptic activity with the epileptogenic zone, typically seen in the interictal state as a region of reduced radiotracer uptake, i.e., hypometabolism [5, 6]. The sensitivity for focus detection in temporal lobe epilepsy (TLE) is reported at 85–90%, generally higher than that for extra-TLE seizures (average 55%, range 45–92%) [7]. Single-photon emission computed tomography (SPECT), another imaging technique for epilepsy, measures cerebral blood flow (CBF) related to the electrical status of the brain [5]. While the sensitivity of interictal SPECT is relatively low, subtraction of interictal from ictal SPECT (SISCOM) can often correctly detect localized ictal hyperperfusion in the epileptogenic focus in TLE (70–90%) and extra-TLE (66%) [7].

Beyond its role in seizure focus detection, advanced nuclear imaging may provide further useful information in epilepsy by targeting specific molecular structures or brain functions. Efforts have been made to explore the potential utilities of imaging for specific PET ligands in epilepsy [8], starting with γ-aminobutyric acid (GABA) and glutamate receptors, which have inhibitory and excitatory functions, respectively.

The central benzodiazepine site on the γ-aminobutyric acid (GABA) receptors can be visualized by ^{11}C-flumazenil administration. In focal epilepsy, ^{11}C-flumazenil PET can be used to localize or lateralize the seizure focus by detecting reduced numbers of benzodiazepine receptors within the epileptogenic region [9]. Compared with ^{18}F-FDG-PET, which usually shows extended hypometabolism beyond the focus in TLE [5], ^{11}C-flumazenil PET may identify smaller, more focused areas of decreased tracer accumulation [10]. ^{18}F-flumazenil PET and ^{123}I-iomazenil SPECT were also developed for benzodiazepine receptor imaging in epilepsy [11, 12].

Glutamate is a major excitatory neurotransmitter in the brain; it makes sense then that chronic neuroinflammation in epilepsy may cause upregulation of glutamate receptors, extracellular accumulation of glutamate, and activation of astrocytes for glutamate clearance [13–15]. Recent advances in molecular imaging have allowed us to visualize glutamate receptors in vivo [16, 17].

^{18}F-GE-179 is a radiotracer for *N*-methyl-D-aspartate (NMDA) receptors, which play crucial roles in various brain functions including memory and learning [16]. Regarding the application of this radiotracer to epilepsy, another study reported focal increases of the distribution volumes of ^{18}F-GE-179 in epilepsy patients as well as a global reduction of the distribution volumes in the epilepsy patients who were taking antidepressants [18]. More recently, ^{11}C-K-2 was developed as a new radiotracer for glutamate alpha-amino-3-hydroxy-5-methyl-4-isoxazole propionic acid (AMPA) receptors, which are thought to underlie several neuropsychiatric disorders [17]. Increased ^{11}C-K-2 uptakes were observed in the epileptogenic foci of temporal lobe epilepsy cases, and a close correlation with the local AMPA receptor protein distribution was revealed in surgical specimens [17].

Monoamine neurotransmitters, such as serotonin or dopamine, are also the targets of nuclear neuroimaging in epilepsy. Serotonin receptor imaging studies using ^{18}F-MPPF or ^{18}F-FCWAY reported reduced uptakes around the focal areas in TLE [19, 20]. It is also reported that patients with epilepsy and depression showed insular reduction of serotonin transporters [21]. Dopamine receptor imaging demonstrated the involvement of basal ganglia in TLE [22]; such findings are expected to lead to elucidation of the roles of dopamine in epilepsy [23].

[11]C-α-Methyl-L-tryptophan (AMT) measures tryptophan metabolism. Interestingly, AMT shows increased uptake in the epileptogenic area interictally, in contrast to other tracers which typically show reduced uptake. In terms of focal AMT uptake, unlike MRI-negative patients with focal cortical dysplasia, patients with normal pathology and intractable epilepsy are prone to surgical failure. AMT is useful in patients with tuberous sclerosis and multiple cortical tubers. However, MRI and ^{18}F-FDG-PET cannot identify which tubers are epileptogenic [24]. ^{18}F-FDG-PET shows reduced uptake in all tubers, reflecting abnormal tissues. Ictal SPECT would be helpful in this regard, but seizures in children are often brief and difficult to capture. It may be that AMT selectively accumulates in epileptogenic tubers; however, the underlying mechanism is unknown, and the evidence is still insufficient.

Supporting evidence shows the involvement of brain inflammation in the process of epileptogenesis [25, 26]. Several specific radiotracers are available to investigate neuroinflammation and epilepsy [8]. PET tracers targeting 18-kDa translocator protein (TSPO), which is located in the outer mitochondrial membrane, can be used to measure neuroinflammation, since immune cells in the brain (including microglia) have high TSPO expression, and TSPO is also thought to be overexpressed during the neuroinflammatory process [27, 28]. ^{11}C-PK11195, which is a first-generation radiotracer targeting TSPO, showed increased binding in Rasmussen's encephalitis [29] and focal cortical dysplasia [30]. Another study using two different TSPO ligands (^{11}C-PBR28 and ^{11}C-DPA-713) reported involvement of the contralateral side in the increased inflammation observed in temporal lobe epilepsy, suggesting ongoing inflammation and a potential role of anti-inflammatory therapy [31]. TSPO PET has thus been providing strong evidence concerning neuroinflammation in epilepsy. The current limitation of TSPO PET is related to the tracers, especially their low-specificity binding (e.g., binding to astrocytes) [32].

More recent studies on epilepsy have focused on the blood–brain barrier (BBB) function and glutamate receptors. Disruption of the BBB is one of the candidate mechanisms underlying epileptogenesis and drug resistance in association with upregulated P-glycoprotein (P-gp; the "permeability" glycoprotein) [33–35]. It is suggested that P-gp plays a role in the BBB's efflux system of antiepileptic drugs, which may be closely related to the development of intractability [36, 37]. R-^{11}C-verapamil PET has been used to measure P-gp and detect BBB dysfunction [38]. Feldmann et al. reported higher baseline P-gp activity in patients with drug-resistant temporal lobe epilepsy compared to seizure-free patients, supporting the relationship between P-gp overactivity and intractability of seizures [39]. More recently, overactivity of P-gp was also confirmed in patients with epileptogenic developmental lesions, which suggests BBB P-gp abnormalities regardless of lesion type [40].

Other studies have also targeted acetylcholine receptors [41] and cannabinoid receptors [42]. In the following sections, we introduce opioid receptor imaging and the impetus for its use in epilepsy, describe in detail its methodology, and identify potential confounders and limitations [43].

2   Materials

The prototype opioid, i.e., morphine, was first isolated in the nineteenth century; thereafter, efforts were made to determine its structure and establish a practical synthetic method [44]. The chemical structures of several endogenous opioid peptides and receptor ligands were identified, and it was also revealed that the endogenous opioids bind to at least three different receptors, i.e., μ-, δ-, and κ-receptors [44]; later, the opioid-receptor-like receptor 1 (ORL1) was discovered [45].

Opioid signaling is involved in pain modulation, reward/addiction, learning/memory, emotion regulation, and epilepsy [46, 47]. Several PET ligands have been developed to visualize opioid systems and functions in the brain in vivo [44]. Of those, the most commonly used tracers are ^{11}C-diprenorphin (DPN) and ^{11}C-carfentanil (CFN). While ^{11}C-DPN is a nonselective opioid receptor antagonist [44], ^{11}C-CFN is a selective agonist of the μ-opioid receptor subtype 1 [48]. Meanwhile, δ, κ, and ORL1 receptors in the human brain have not yet been well studied [44].

The μ-opioid receptor mediates endogenous β-endorphins, endomorphins, enkephalins, and many exogenous opioid agonists. While its significant role in pain modulation is well known [48], recent studies have reported associations between μ-opioid receptors and various neuropsychiatric disorders, including addiction, affective disorders, and schizophrenia [49–53]. In epilepsy, it is known that endogenous opioids are released during spontaneous and provoked seizures and may be involved in seizure termination [47, 54]. This ictal opioid release leads to a subsequent hyperexpression of opioid receptors [54]. These adaptive processes start within hours after seizures and are likely to persist for several days. Postictal and interictal changes of opioid transmission might lead to affective dysregulation as reflected in impaired emotional processing in temporal lobe epilepsy patients [55, 56]. Thus, it is reasonable to investigate the relationships between epilepsy, affective symptoms, and μ-opioid systems.

3 Methods

The highly potent opioid carfentanil, also known as 4-carbomethoxy fentanyl, can be synthesized from its carboxylate precursor [57]. Participants undergo the intravenous injection of ^{11}C-CFN (dose range: 110–330 MBq) after a low-dose CT scan for attenuation correction. The subsequent PET/CT emission scan duration is 90 min in 26 frames (8×15 s, 3×60 s, 5×120 s, 5×300 s, and 5×600 s, to a total of 5400 s). For spatial normalization of the individual's scan, participants also undergo a 3D volumetric T1-weighted structural brain MRI scan with an isotropic resolution of 1 mm.

The dynamic PET images are initially processed by attenuation correction and motion correction using frame-by-frame realignment. When there are gross movements of participants during the scan, the affected frames are typically excluded. After spatial coregistration to the structural MRI, the nonlinear deformation parameters derived from unified segmentation of MRI are applied to the PET images. Then, the nondisplaceable binding potential (BP) values of ^{11}C-CFN can be quantified by the simplified reference tissue model (SRTM) with occipital lobe grey matter as the reference [51, 58]. To avoid invasive arterial blood sampling procedures during the scan, it is necessary to correct the tissue BP images by the BP of an appropriate reference region where no specific uptake of the PET ligand is observed. In the case of ^{11}C-CFN, the standard reference region is the occipital lobe gray matter. A sample of the normalized ^{11}C-CFN image is shown in Fig. 1. For these processes, we utilize MIAKAT™ software (www. miakat.org) and statistical parametric mapping software 12 (SPM12; http://www.fil.ion.ucl.ac.uk/spm/).

For the analysis of the normalized PET images, there are mainly two methods, i.e., voxel-based and region-of-interest (ROI) analyses. Voxel-based analysis (VBA) involves a voxel-wise comparison or correlation, which can be used for whole-brain analyses without any specific ROI settings [59]; it is useful for exploratory research tasks because of this whole-brain approach. On the other hand, in ROI analysis, specific ROIs are established, and the values, e.g., of BP of ^{11}C-CFN, within those regions are calculated. Since it is possible to obtain numerical values for each subject at the individual level, these numerical values can be used as indicators in clinical

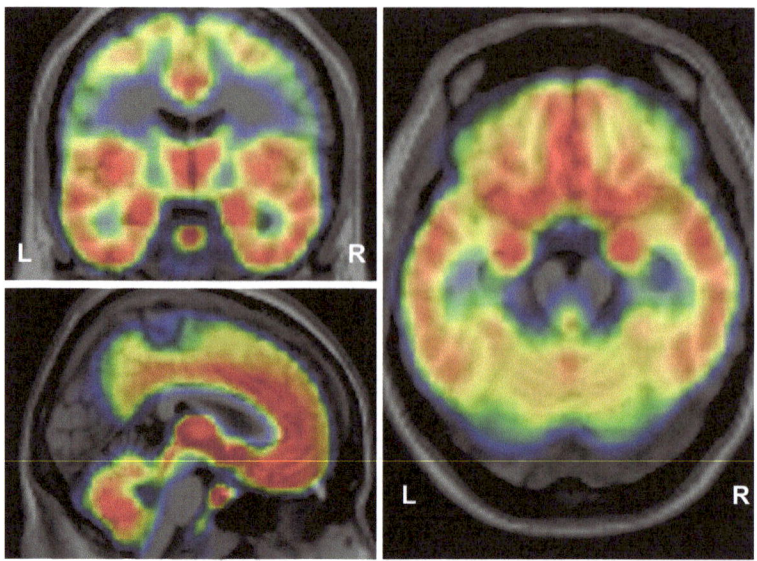

Fig. 1 A sample normalized ^{11}C-CFN image

practice, e.g., as cut-off values for diagnosis. ROI analysis can be used when it is possible to set relevant, reliable ROIs from previous literature: in the case of TLE and affective disorders, the limbic regions such as the hippocampus and amygdala and the prefrontal regions are good candidates for ROI analysis. Sample images of both types of ^{11}C-CFN imaging are shown in Fig. 2a (voxel-based group comparison) and Fig. 2b (ROI analysis).

4   Points of Attention

Unlike neurodegenerative diseases, epileptic seizures cause dynamic changes in brain function so that we have to pay attention to the timing of the scan and seizures. The opioid system is no exception, with the peri-ictal period resulting in the release of opioid-like substances and a decrease in receptor-binding capacity. Therefore, the time interval between the scan and the last seizure should be carefully monitored. For image processing, such as spatial standardization, automated analysis methods have been established and are usually used, but sometimes errors can cause misalignments and other problems. If large outliers are produced, they can lead to erroneous conclusions about research results; therefore, all images should be checked visually and not simply data extracted. As a general note on PET studies, clinical PET tracers need to have acceptable brain penetration, low nonspecific bindings, high affinity, and target selectivity [8]. Also, attention should be paid to off-target binding, which is the uptake in structures unrelated to the molecule being targeted.

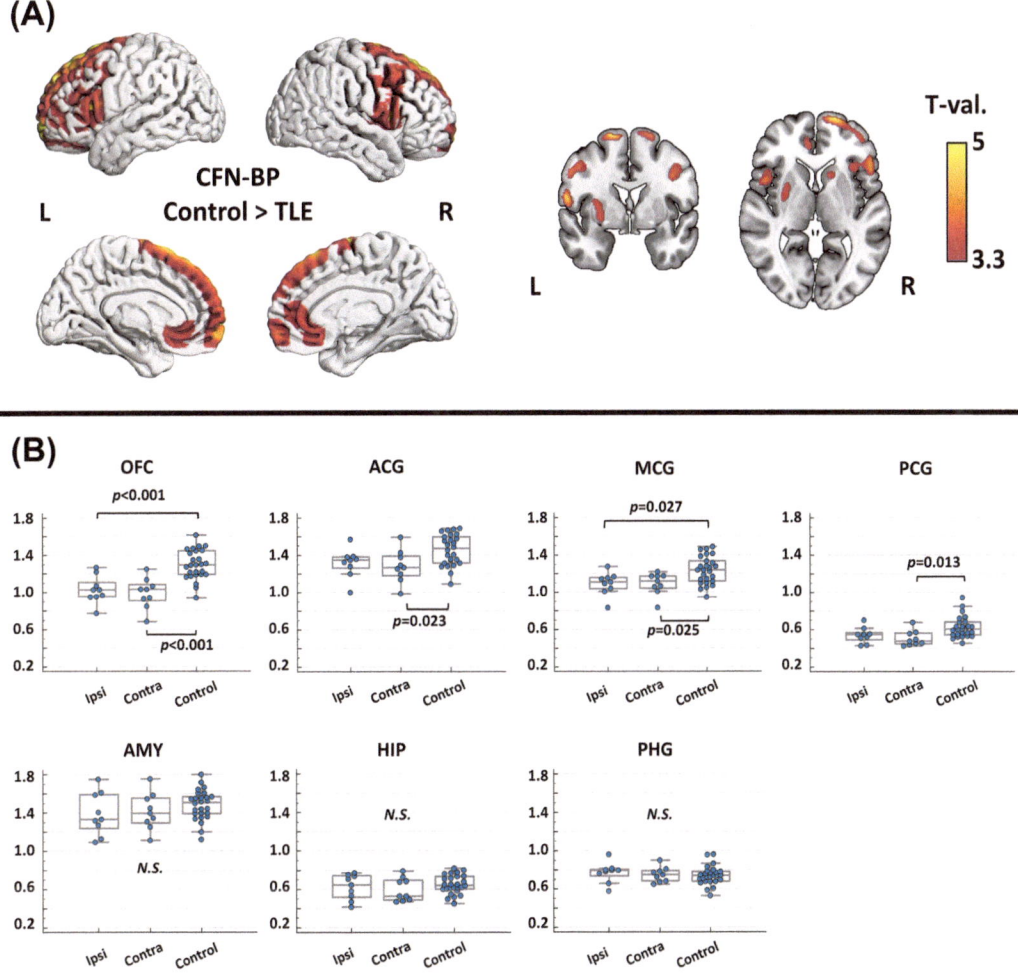

Fig. 2 Results of voxel-based group comparison (**a**) and ROI analysis (**b**). (Modified from ref. 43)

References

1. Fisher RS, van Emde BW, Blume W, Elger C, Genton P, Lee P et al (2005) Epileptic seizures and epilepsy: definitions proposed by the International League Against Epilepsy (ILAE) and the International Bureau for Epilepsy (IBE). Epilepsia 46(4):470–472. https://doi.org/10.1111/j.0013-9580.2005.66104.x. PubMed PMID: 15816939

2. Collaborators GBDE (2019) Global, regional, and national burden of epilepsy, 1990–2016: a systematic analysis for the Global Burden of Disease Study 2016. Lancet Neurol 18(4): 357–375. https://doi.org/10.1016/S1474-4422(18)30454-X. Epub 2019/02/19. PubMed PMID: 30773428; PubMed Central PMCID: PMCPMC6416168

3. Chen Z, Brodie MJ, Liew D, Kwan P (2018) Treatment outcomes in patients with newly diagnosed epilepsy treated with established and new antiepileptic drugs: a 30-Year Longitudinal Cohort Study. JAMA Neurol 75(3): 279–286. https://doi.org/10.1001/jamaneurol.2017.3949. PubMed PMID: 29279892; PubMed Central PMCID: PMCPMC5885858

4. Rathore C, Radhakrishnan K (2015) Concept of epilepsy surgery and presurgical evaluation. Epileptic Disord 17(1):19–31. https://doi.org/10.1684/epd.2014.0720;quiz PubMed PMID: 25652945

5. Kumar A, Chugani HT (2013) The role of radionuclide imaging in epilepsy, part 1:

sporadic temporal and extratemporal lobe epilepsy. J Nucl Med 54(10):1775–1781. https://doi.org/10.2967/jnumed.112.114397. PubMed PMID: 23970368

6. Sone D, Maikusa N, Sato N, Kimura Y, Ota M, Matsuda H (2019) Similar and differing distributions between ^{18}F-FDG-PET and arterial spin Labeling imaging in temporal lobe epilepsy. Front Neurol 10:318. https://doi.org/10.3389/fneur.2019.00318

7. Shigemoto Y, Sone D, Kimura Y, Sato N, Matsuda H (2020) Nuclear imaging in epilepsy: principles and Progress. Epilepsy Seizure 12(1):40–48. https://doi.org/10.3805/eands.12.40

8. Galovic M, Koepp M (2016) Advances of molecular imaging in epilepsy. Curr Neurol Neurosci Rep 16(6):58. https://doi.org/10.1007/s11910-016-0660-7. Epub 2016/04/27. PubMed PMID: 27113252; PubMed Central PMCID: PMCPMC4844640

9. la Fougere C, Rominger A, Forster S, Geisler J, Bartenstein P (2009) PET and SPECT in epilepsy: a critical review. Epilepsy Behav 15(1):50–55. https://doi.org/10.1016/j.yebeh.2009.02.025. Epub 2009/02/21. PubMed PMID: 19236949

10. Ryvlin P, Bouvard S, Le Bars D, De Lamerie G, Gregoire MC, Kahane P et al (1998) Clinical utility of flumazenil-PET versus ^{18}F]fluorodeoxyglucose-PET and MRI in refractory partial epilepsy. A prospective study in 100 patients. Brain 121(Pt 11):2067–2081. PubMed PMID: 9827767

11. Lamusuo S, Ruottinen HM, Knuuti J, Harkonen R, Ruotsalainen U, Bergman J et al (1997) Comparison of ^{18}F]FDG-PET, [99mTc]-HMPAO-SPECT, and [123I]-iomazenil-SPECT in localising the epileptogenic cortex. J Neurol Neurosurg Psychiatry 63(6):743–748. PubMed PMID: 9416808; PubMed Central PMCID: PMC2169853

12. Vivash L, Gregoire MC, Lau EW, Ware RE, Binns D, Roselt P et al (2013) ^{18}F-flumazenil: a gamma aminobutyric acid A-specific PET radiotracer for the localization of drug-resistant temporal lobe epilepsy. J Nucl Med 54(8):1270–1277. https://doi.org/10.2967/jnumed.112.107359. Epub 2013/07/15. PubMed PMID: 23857513

13. Eid T, Thomas MJ, Spencer DD, Runden-Pran E, Lai JC, Malthankar GV et al (2004) Loss of glutamine synthetase in the human epileptogenic hippocampus: possible mechanism for raised extracellular glutamate in mesial temporal lobe epilepsy. Lancet 363(9402):28–37. https://doi.org/10.1016/s0140-6736(03)15166-5. Epub 2004/01/16. PubMed PMID: 14723991

14. Pan JW, Williamson A, Cavus I, Hetherington HP, Zaveri H, Petroff OA et al (2008) Neurometabolism in human epilepsy. Epilepsia 49 (Suppl 3):31–41. https://doi.org/10.1111/j.1528-1167.2008.01508.x. Epub 2008/05/28. PubMed PMID: 18304254; PubMed Central PMCID: PMCPMC3657747

15. Sharma AA, Szaflarski JP (2020) In vivo imaging of neuroinflammatory targets in treatment-resistant epilepsy. Curr Neurol Neurosci Rep 20(4):5. https://doi.org/10.1007/s11910-020-1025-9. Epub 2020/03/14. PubMed PMID: 32166626

16. McGinnity CJ, Hammers A, Riano Barros DA, Luthra SK, Jones PA, Trigg W et al (2014) Initial evaluation of ^{18}F-GE-179, a putative PET tracer for activated N-methyl D-aspartate receptors. J Nucl Med 55(3):423–430. https://doi.org/10.2967/jnumed.113.130641. Epub 2014/02/15. PubMed PMID: 24525206

17. Miyazaki T, Nakajima W, Hatano M, Shibata Y, Kuroki Y, Arisawa T et al (2020) Visualization of AMPA receptors in living human brain with positron emission tomography. Nat Med 26(2):281–288. https://doi.org/10.1038/s41591-019-0723-9. Epub 2020/01/22. PubMed PMID: 31959988

18. McGinnity CJ, Koepp MJ, Hammers A, Riano Barros DA, Pressler RM, Luthra S et al (2015) NMDA receptor binding in focal epilepsies. J Neurol Neurosurg Psychiatry 86(10):1150–1157. https://doi.org/10.1136/jnnp-2014-309897. Epub 2015/05/21. PubMed PMID: 25991402; PubMed Central PMCID: PMCPMC4602274

19. Didelot A, Ryvlin P, Lothe A, Merlet I, Hammers A, Mauguiere F (2008) PET imaging of brain 5-HT1A receptors in the preoperative evaluation of temporal lobe epilepsy. Brain J Neurol 131(Pt 10):2751–2764. https://doi.org/10.1093/brain/awn220. PubMed PMID: 18790822

20. Liew CJ, Lim YM, Bonwetsch R, Shamim S, Sato S, Reeves-Tyer P et al (2009) ^{18}F-FCWAY and ^{18}F-FDG PET in MRI-negative temporal lobe epilepsy. Epilepsia 50(2):234–239. https://doi.org/10.1111/j.1528-1167.2008.01789.x. PubMed PMID: 18801033; PubMed Central PMCID: PMCPMC2642908

21. Martinez A, Finegersh A, Cannon DM, Dustin I, Nugent A, Herscovitch P et al (2013) The 5-HT1A receptor and 5-HT transporter in temporal lobe epilepsy. Neurology 80(16):1465–1471. https://doi.org/10.

1212/WNL.0b013e31828cf809.   Epub 2013/03/20. PubMed PMID: 23516322; PubMed   Central   PMCID: PMCPMC3662359

22. Werhahn KJ, Landvogt C, Klimpe S, Buchholz HG, Yakushev I, Siessmeier T et al (2006) Decreased dopamine D2/D3-receptor binding in temporal lobe epilepsy: an [^{18}F]fallypride PET study. Epilepsia 47(8):1392–1396. https://doi.org/10.1111/j.1528-1167.2006. 00561.x. PubMed PMID: 16922886

23. Haut SR, Albin RL (2008) Dopamine and epilepsy: hints of complex subcortical roles. Neurology 71(11):784–785. https://doi.org/10. 1212/01.wnl.0000325637.38931.27. PubMed PMID: 18779507

24. Chugani HT, Luat AF, Kumar A, Govindan R, Pawlik K, Asano E (2013) Alpha-[^{11}C]-Methyl-L-tryptophan—PET in 191 patients with tuberous sclerosis complex. Neurology 81(7):674–680. https://doi.org/10.1212/ WNL.0b013e3182a08f3f. PubMed PMID: 23851963; PubMed Central PMCID: PMCPMC3775695

25. Sone D (2023) Contribution of neuroimaging studies to the understanding of immunology and inflammation in epilepsy. In: Rezaei N, Yazdanpanah N (eds) Translational neuroimmunology, vol 7. Academic Press, Cambridge, MA, pp 411–423

26. Vezzani A, Friedman A, Dingledine RJ (2013) The role of inflammation in epileptogenesis. Neuropharmacology 69:16–24. https://doi. org/10.1016/j.neuropharm.2012.04.004. Epub   2012/04/24.   PubMed   PMID: 22521336; PubMed Central PMCID: PMCPMC3447120

27. Owen DR, Guo Q, Kalk NJ, Colasanti A, Kalogiannopoulou D, Dimber R et al (2014) Determination of [(11)C]PBR28 binding potential in vivo: a first human TSPO blocking study. J Cereb Blood Flow Metab 34(6): 989–994. https://doi.org/10.1038/jcbfm. 2014.46.   Epub   2014/03/20. PubMed PMID: 24643083; PubMed Central PMCID: PMCPMC4050243

28. Vivash L, O'Brien TJ (2016) Imaging microglial activation with TSPO PET: lighting up neurologic diseases? J Nucl Med 57(2): 165–168. https://doi.org/10.2967/jnumed. 114.141713.   Epub   2015/12/25. PubMed PMID: 26697963

29. Banati RB, Goerres GW, Myers R, Gunn RN, Turkheimer FE, Kreutzberg GW et al (1999) [^{11}C](R)-PK11195 positron emission tomography imaging of activated microglia in vivo in Rasmussen's encephalitis. Neurology 53(9): 2199–2203. https://doi.org/10.1212/wnl.

53.9.2199.   Epub   1999/12/22. PubMed PMID: 10599809

30. Butler T, Ichise M, Teich AF, Gerard E, Osborne J, French J et al (2013) Imaging inflammation in a patient with epilepsy due to focal cortical dysplasia. J Neuroimaging 23(1): 129–131.   https://doi.org/10.1111/j. 1552-6569.2010.00572.x. Epub 2011/01/ 13. PubMed PMID: 21223436; PubMed Central PMCID: PMCPMC5303618

31. Gershen LD, Zanotti-Fregonara P, Dustin IH, Liow JS, Hirvonen J, Kreisl WC et al (2015) Neuroinflammation in temporal lobe epilepsy measured using positron emission tomographic imaging of translocator protein. JAMA Neurol 72(8):882–888. https://doi. org/10.1001/jamaneurol.2015.0941. Epub 2015/06/09. PubMed PMID: 26052981; PubMed   Central   PMCID: PMCPMC5234042

32. Cosenza-Nashat M, Zhao ML, Suh HS, Morgan J, Natividad R, Morgello S et al (2009) Expression of the translocator protein of 18 kDa by microglia, macrophages and astrocytes based on immunohistochemical localization in abnormal human brain. Neuropathol Appl Neurobiol 35(3):306–328. https://doi.org/10.1111/j.1365-2990.2008. 01006.x. Epub 2008/12/17. PubMed PMID: 19077109;   PubMed   Central   PMCID: PMCPMC2693902

33. Tishler DM, Weinberg KI, Hinton DR, Barbaro N, Annett GM, Raffel C (1995) MDR1 gene expression in brain of patients with medically intractable epilepsy. Epilepsia 36(1):1–6. https://doi.org/10.1111/j. 1528-1157.1995.tb01657.x. Epub 1995/ 01/01. PubMed PMID: 8001500

34. Tomkins O, Shelef I, Kaizerman I, Eliushin A, Afawi Z, Misk A et al (2008) Blood-brain barrier disruption in post-traumatic epilepsy. J Neurol Neurosurg Psychiatry 79(7):774–777. https://doi.org/10.1136/jnnp.2007. 126425. Epub 2007/11/10. PubMed PMID: 17991703

35. Stepien KM, Tomaszewski M, Tomaszewska J, Czuczwar SJ (2012) The multidrug transporter P-glycoprotein in pharmacoresistance to antiepileptic drugs. Pharmacol Rep 64(5): 1011–1019. https://doi.org/10.1016/ s1734-1140(12)70900-3. Epub 2012/12/ 15. PubMed PMID: 23238460

36. Loscher W (2007) Drug transporters in the epileptic brain. Epilepsia 48(Suppl 1):8–13. https://doi.org/10.1111/j.1528-1167.2007. 00993.x. Epub 2007/02/24. PubMed PMID: 17316407

37. Potschka H (2012) Role of CNS efflux drug transporters in antiepileptic drug delivery: overcoming CNS efflux drug transport. Adv Drug Deliv Rev 64(10):943–952. https://doi.org/10.1016/j.addr.2011.12.007. Epub 2012/01/03. PubMed PMID: 22210135

38. Luurtsema G, Molthoff CF, Schuit RC, Windhorst AD, Lammertsma AA, Franssen EJ (2005) Evaluation of (R)-[^{11}C]verapamil as PET tracer of P-glycoprotein function in the blood-brain barrier: kinetics and metabolism in the rat. Nucl Med Biol 32(1):87–93. https://doi.org/10.1016/j.nucmedbio.2004.06.007. Epub 2005/02/05. PubMed PMID: 15691665

39. Feldmann M, Asselin MC, Liu J, Wang S, McMahon A, Anton-Rodriguez J et al (2013) P-glycoprotein expression and function in patients with temporal lobe epilepsy: a case-control study. Lancet Neurol 12(8):777–785. https://doi.org/10.1016/S1474-4422(13)70109-1. Epub 2013/06/22. PubMed PMID: 23786896

40. Ilyas-Feldmann M, Asselin MC, Wang S, McMahon A, Anton-Rodriguez J, Brown G et al (2020) P-glycoprotein overactivity in epileptogenic developmental lesions measured in vivo using (R)-[(11) C]verapamil PET. Epilepsia 61(7):1472–1480. https://doi.org/10.1111/epi.16581. Epub 2020/07/07. PubMed PMID: 32627849

41. Picard F, Bruel D, Servent D, Saba W, Fruchart-Gaillard C, Schollhorn-Peyronneau MA et al (2006) Alteration of the in vivo nicotinic receptor density in ADNFLE patients: a PET study. Brain J Neurol 129(Pt 8):2047–2060. https://doi.org/10.1093/brain/awl156. Epub 20060630. PubMed PMID: 16815873

42. Goffin K, Van Paesschen W, Van Laere K (2011) In vivo activation of endocannabinoid system in temporal lobe epilepsy with hippocampal sclerosis. Brain 134(Pt 4):1033–1040. https://doi.org/10.1093/brain/awq385. Epub 20110208. PubMed PMID: 21303859

43. Sone D, Galovic M, Myers J, Leonhardt G, Rabiner I, Duncan JS et al (2023) Contribution of the mu-opioid receptor system to affective disorders in temporal lobe epilepsy: a bidirectional relationship? Epilepsia 64(2):420–429. https://doi.org/10.1111/epi.17463. Epub 20221228. PubMed PMID: 36377838

44. Cumming P, Marton J, Lilius TO, Olberg DE, Rominger A (2019) A survey of molecular imaging of opioid receptors. Molecules 24(22). https://doi.org/10.3390/molecules24224190. Epub 20191119.

PubMed PMID: 31752279; PubMed Central PMCID: PMCPMC6891617

45. Meunier JC (1997) Nociceptin/orphanin FQ and the opioid receptor-like ORL1 receptor. Eur J Pharmacol 340(1):1–15. https://doi.org/10.1016/s0014-2999(97)01411-8. PubMed PMID: 9527501

46. Valentino RJ, Volkow ND (2018) Untangling the complexity of opioid receptor function. Neuropsychopharmacology 43(13):2514–2520. https://doi.org/10.1038/s41386-018-0225-3. Epub 20180924. PubMed PMID: 30250308; PubMed Central PMCID: PMCPMC6224460

47. Koepp MJ, Richardson MP, Brooks DJ, Duncan JS (1998) Focal cortical release of endogenous opioids during reading-induced seizures. Lancet 352(9132):952–955. PubMed PMID: 9752818

48. Eriksson O, Antoni G (2015) [^{11}C]Carfentanil binds preferentially to mu-opioid receptor subtype 1 compared to subtype 2. Mol Imaging 14:476–483. PubMed PMID: 26461068

49. Hsu DT, Sanford BJ, Meyers KK, Love TM, Hazlett KE, Walker SJ et al (2015) It still hurts: altered endogenous opioid activity in the brain during social rejection and acceptance in major depressive disorder. Mol Psychiatry 20(2):193–200. https://doi.org/10.1038/mp.2014.185. PubMed PMID: 25600108; PubMed Central PMCID: PMCPMC4469367

50. Mick I, Myers J, Ramos AC, Stokes PR, Erritzoe D, Colasanti A et al (2016) Blunted endogenous opioid release following an oral amphetamine challenge in pathological gamblers. Neuropsychopharmacology 41(7):1742–1750. https://doi.org/10.1038/npp.2015.340. Epub 2015/11/11. PubMed PMID: 26552847; PubMed Central PMCID: PMCPMC4869041

51. Ashok AH, Myers J, Reis Marques T, Rabiner EA, Howes OD (2019) Reduced mu opioid receptor availability in schizophrenia revealed with [(11)C]-carfentanil positron emission tomographic imaging. Nat Commun 10(1):4493. https://doi.org/10.1038/s41467-019-12366-4. Epub 2019/10/05. PubMed PMID: 31582737; PubMed Central PMCID: PMCPMC6776653

52. Nummenmaa L, Karjalainen T, Isojarvi J, Kantonen T, Tuisku J, Kaasinen V et al (2020) Lowered endogenous mu-opioid receptor availability in subclinical depression and anxiety. Neuropsychopharmacology 45(11):1953–1959. https://doi.org/10.1038/s41386-020-0725-9. Epub 2020/05/31.

PubMed PMID: 32473595; PubMed Central PMCID: PMCPMC7608336

53. Turton S, Myers JF, Mick I, Colasanti A, Venkataraman A, Durant C et al (2020) Blunted endogenous opioid release following an oral dexamphetamine challenge in abstinent alcohol-dependent individuals. Mol Psychiatry 25(8):1749–1758. https://doi.org/10.1038/s41380-018-0107-4. Epub 2018/06/27. PubMed PMID: 29942043; PubMed Central PMCID: PMCPMC6169731

54. Hammers A, Asselin MC, Hinz R, Kitchen I, Brooks DJ, Duncan JS et al (2007) Upregulation of opioid receptor binding following spontaneous epileptic seizures. Brain 130(Pt 4): 1009–1016. https://doi.org/10.1093/brain/awm012. PubMed PMID: 17301080

55. Meletti S, Benuzzi F, Rubboli G, Cantalupo G, Stanzani Maserati M, Nichelli P et al (2003) Impaired facial emotion recognition in early-onset right mesial temporal lobe epilepsy. Neurology 60(3):426–431. PubMed PMID: 12578923

56. Batut AC, Gounot D, Namer IJ, Hirsch E, Kehrli P, Metz-Lutz MN (2006) Neural responses associated with positive and negative emotion processing in patients with left versus right temporal lobe epilepsy. Epilepsy Behavior 9(3):415–423. https://doi.org/10.1016/j.yebeh.2006.07.013. PubMed PMID: 16949873

57. Dannals RF, Ravert HT, Frost JJ, Wilson AA, Burns HD, Wagner HN Jr (1985) Radiosynthesis of an opiate receptor binding radiotracer: [11C]carfentanil. Int J Appl Radiat Isot 36(4): 303–306. https://doi.org/10.1016/0020-708x(85)90089-4. PubMed PMID: 2991142

58. Lammertsma AA, Hume SP (1996) Simplified reference tissue model for PET receptor studies. NeuroImage 4(3 Pt 1):153–158. https://doi.org/10.1006/nimg.1996.0066. Epub 1996/12/01. PubMed PMID: 9345505

59. Ashburner J, Friston KJ (2000) Voxel-based morphometry—the methods. NeuroImage 11(6 Pt 1):805–821. https://doi.org/10.1006/nimg.2000.0582. PubMed PMID: 10860804

Chapter 8

High-Affinity State of Dopamine $D_{2/3}$ Receptors in Schizophrenia: A Dual-Radioligand Approach

Fumitoshi Kodaka

Abstract

Dopamine (DA) $D_{2/3}R$ receptors ($D_{2/3}Rs$) are of central interest in studies of the pathophysiology of schizophrenia. The sensitivity of $D_{2/3}Rs$ to DA can differ depending on the affinity state of $D_{2/3}Rs$ for DA. Early in vitro studies reported that $D_{2/3}Rs$ exist in two interconvertible affinity states for DA, referred to as high-affinity ($D_{2/3}^{high}$) and low-affinity ($D_{2/3}^{low}$) states. An increased proportion of $D_{2/3}^{high}$ is assumed to correlate with dopamine hypersensitivity, suggesting a relationship with the psychotic symptoms of schizophrenia. Recently, several $D_{2/3}R$ agonist radioligands that are thought to bind preferentially to $D_{2/3}^{high}$ in the living human brain have been developed. However, there are controversies regarding whether $D_{2/3}R$ agonist radioligands can be used to quantify only $D_{2/3}^{high}$ in vivo. The findings of PET studies with $D_{2/3}R$ agonist radioligands indicate that various physiological conditions can modulate the affinity states of $D_{2/3}Rs$. Given that the ratio of $D_{2/3}^{high}$ to $D_{2/3}^{low}$ fluctuates in various physiological states, a dual-radioligand approach using $D_{2/3}R$ agonist radioligands and a $D_{2/3}R$ antagonist radioligand to estimate the ratio of $D_{2/3}^{high}$ to $D_{2/3}^{low}$ is needed to detect alterations in $D_{2/3}^{high}$ in schizophrenia patients.

Key words Molecular imaging, Schizophrenia, Affinity states of dopamine receptors, Antipsychotics, Receptor occupancy

1 Introduction

Dopamine (DA) $D_{2/3}R$ receptors ($D_{2/3}Rs$) are of central interest in studies of the pathophysiology of schizophrenia. Early research on the pathophysiology of schizophrenia using positron emission tomography (PET) focused on the striatum because of the higher $D_{2/3}R$ density in this region than in cortical regions. Early post-mortem studies have reported a greater density of $D_{2/3}Rs$ in the striatum in patients with schizophrenia than in healthy individuals [1], yet subsequent PET and single photon emission computed tomography (SPECT) studies of drug-free and/or drug-naive patients with schizophrenia have revealed inconclusive evidence for striatal $D_{2/3}R$ binding when compared to healthy individuals. Several meta-analyses of PET and SPECT studies have found that

Daichi Sone (ed.), *Molecular Imaging for Brain Diseases*, Neuromethods, vol. 222, https://doi.org/10.1007/978-1-0716-4494-2_8,

patients with schizophrenia exhibit slightly greater striatal $D_{2/3}R$ binding than healthy individuals [2, 3].

Psychotic symptoms have been reported to worsen with the use of psychostimulants (i.e., amphetamine) in patients with schizophrenia [4, 5], leading to increased stimulation of $D_{2/3}Rs$ due to excessive presynaptic DA release. An alternative mechanism is postulated for the increased density of $D_{2/3}Rs$ and/or the hypersensitivity of $D_{2/3}Rs$ to DA. However, a previous PET study reported that the increase in $D_{2/3}R$ binding in patients with schizophrenia is approximately 20% [6]. Because the decrease in striatal $D_{2/3}R$ density with age is approximately 6–13% by decade [7, 8], a change in $D_{2/3}R$ density of 20% may not explain psychotic symptoms.

The sensitivity of $D_{2/3}Rs$ to DA can be modulated by the state of $D_{2/3}Rs$ with high affinity for DA. Early in vitro studies reported that $D_{2/3}Rs$ exist in two interconvertible affinity states for DA, referred to as high-affinity ($D_{2/3}^{high}$) and low-affinity ($D_{2/3}^{low}$) states [9–12]. An increased proportion of $D_{2/3}^{high}$ is assumed to correlate with dopamine hypersensitivity, suggesting a relationship with the psychotic symptoms of schizophrenia. In vitro competition binding assay studies using homogenized striatal tissue reported that the proportion of $D_{2/3}^{high}$ is approximately 50% [13, 14]. Although $D_{2/3}^{high}$ $D_{2/3}R$ agonist ligands have been developed, $D_{2/3}^{high}$ has yet to be robustly quantified.

2   Materials

2.1   Affinity States of Dopamine $D_{2/3}$ Receptors

$D_{2/3}^{high}$ and $D_{2/3}^{low}$ can be observed through competition binding assays. Both high- and low-affinity fractions are observed as *biphasic* curves when an agonist to an antagonist radioligand for binding to $D_{2/3}Rs$ is used. Only the high-affinity fraction is observed as a *monophasic* curve when an agonist to an agonist radioligand for binding to $D_{2/3}Rs$ is used [9]. Previous assays revealed that agonists bind to $D_{2/3}Rs$ with high and low affinity in the absence of guanine nucleotide triphosphate (GTP) and only with low affinity in the presence of GTP. Because GTP binds to the $G\alpha$ subunit in the G protein, the activated G protein dissociates from $D_{2/3}Rs$, leading to the low affinity of $D_{2/3}Rs$ for the agonist.

In the agonist-receptor-G protein complex model (i.e., the ternary complex model), $D_{2/3}Rs$ exist in two interconvertible affinity states. In the G-protein-coupled high-affinity state, the G protein binds preferentially to agonists, whereas in the G-protein uncoupled low-affinity state, the G protein does not bind to antagonists. The ternary complex model provides the theoretical basis that $D_{2/3}Rs$ exist in both the $D_{2/3}^{high}$ and $D_{2/3}^{low}$ states in the living human brain.

The proportion of $D_{2/3}^{high}$ can be altered by modulation of the dopaminergic system. Rodents sensitized to amphetamine show a reduction in $D_{2/3}R$ binding but a two-to-four-fold increase in the proportion of $D_{2/3}^{high}$. The proportion of $D_{2/3}^{high}$ is also increased with chronic administration of antipsychotics [15–17].

3 Methods

3.1 Quantification of $D_{2/3}^{high}$ in the Human Brain with PET

3.1.1 Quantification of the High-Affinity State of Dopamine $D_{2/3}Rs$ Through Agonist Ligand Binding

In vivo, $D_{2/3}^{high}$ density can be estimated by determining the ratio of the binding of $D_{2/3}R$ agonist ligands to the binding of $D_{2/3}R$ antagonist ligands. A previous study reported that the dissociation constants (K_d) of endogenous DA to $D_{2/3}^{high}$ and $D_{2/3}^{low}$ were 7 nM and 1720 nM, respectively, indicating an approximately 250-fold greater affinity for $D_{2/3}^{high}$ than for $D_{2/3}^{low}$ [10]. Because the K_d values of $D_{2/3}R$ agonists (e.g., (−)-N-propyl-norapomorphine (NPA)) [18, 19] to $D_{2/3}^{high}$ and $D_{2/3}^{low}$ have been reported to be 0.07–0.4 nM and 20–200 nM, respectively [10, 20, 21], $D_{2/3}^{high}$ density can be estimated by measuring the binding of $D_{2/3}R$ agonist radioligands. Because $D_{2/3}R$ antagonists bind to $D_{2/3}^{high}$ and $D_{2/3}^{low}$, the total density of $D_{2/3}^{high}$ and $D_{2/3}^{low}$ can be estimated by measuring the binding of $D_{2/3}R$ antagonist radioligands.

Recently, several $D_{2/3}R$ agonist radioligands that are thought to bind preferentially to $D_{2/3}^{high}$ in the living human brain have been developed, including [^{11}C]NPA [22], the methoxy analogue of NPA; (R)-2-[^{11}C]methoxy-N-n-propylnorapomorphine ([^{11}C] MNPA) [23, 24]; and ^{11}C-4-N-propyl-,3,4a,5,6,10b-hexahydro-2H-naphth[1,2-b][1,4]-oxazin-9-ol (^{11}C-PHNO) [25, 26]. The selectivity for dopamine D_2 receptors (D_2Rs) over dopamine D_3 receptors (D_3Rs) differs among radioligands [27], indicating that these radioligands can be used for research on the pathophysiology of schizophrenia. NPA and MNPA are reported to have almost identical selectivities for D_2Rs and D_3Rs, whereas PHNO is reported to have selectivity for D_3Rs over D_2Rs. The newly developed $D_{2/3}R$ agonist ligand [^{3}H]MCL-536 is reported to have high selectivity for D_2Rs [27].

Previous in vivo PET studies using $D_{2/3}R$ agonist ligands have reported the $D_{2/3}^{high}$ density in primates. In a PET study in baboons using [^{11}C]NPA and the $D_{2/3}R$ antagonist ligand [^{11}C] raclopride to measure the dissociation constant (K_d), the estimated $D_{2/3}^{high}$ and $D_{2/3}^{low}$ densities (B_{max}) were 0.16 ± 0.01 nM and 21.6 ± 2.8 nM, respectively, for [^{11}C]NPA and 1.59 ± 0.01 nM and 27.3 ± 3.9 nM, respectively, for [^{11}C]raclopride. The authors reported that the B_{max} of [^{11}C]NPA was 79% of that of [^{11}C] raclopride, consistent with previous studies [28].

3.1.2 Controversies in PET Quantification of In Vivo $D_{2/3}^{high}$

There are controversies regarding whether $D_{2/3}R$ agonist radioligands can be used to quantify only $D_{2/3}^{high}$ in vivo. The findings of PET studies in animals have indicated that anesthetics, such as isoflurane, sevoflurane, or ketamine, may modulate the affinity states of $D_{2/3}$Rs. Several PET studies with $D_{2/3}$Rs agonist ($[^{11}C]$ NPA, $[^{11}C]$MNPA, and $[^{11}C]$PHNO) and antagonist ($[^{11}C]$raclopride) radioligands in primates demonstrated that agonist radioligands showed almost twofold greater displacement after amphetamine challenge than did the antagonist radioligand [29–31], indicating that agonist radioligands bind preferentially to $D_{2/3}^{high}$. In their PET study of anesthetized baboons, Narendran et al. administered 0.3 mg/kg, 0.5 mg/kg, and 1.0 mg/kg amphetamine to estimate the displacement of $[^{11}C]$ NPA and $[^{11}C]$raclopride binding to $D_{2/3}$Rs by DA; the $[^{11}C]$ raclopride binding potential decreased by 24 ± 10%, 32 ± 6%, and 44 ± 9%, whereas the $[^{11}C]$NPA binding potential decreased by 32 ± 2%, 45 ± 3%, and 53 ± 9%, respectively. These results indicate that $[^{11}C]$NPA is more sensitive to DA release than $[^{11}C]$ raclopride [29]. Other amphetamine challenge PET studies with $[^{11}C]$NPA, $[^{11}C]$PHNO, and $[^{18}F]$MCL demonstrated almost identical results [30–32]. In contrast, other previous PET studies failed to demonstrate a statistically significant decrease in agonist radioligand binding compared with antagonist radioligand binding after amphetamine administration in conscious animals.

The findings of previous studies have suggested that isoflurane and sevoflurane can increase the turnover of endogenous DA in the synaptic cleft, as indicated by increased DA synthesis, increased DA transporter (DAT) binding, and decreased $D_{2/3}$R binding. Although several factors can be involved in greater displacement of agonist radioligands after amphetamine challenge than of antagonist radioligands, these anesthetics may increase the ratio of $D_{2/3}^{high}$ to $D_{2/3}^{low}$ through increased baseline dopamine (tonic) release and the internalization of low-affinity $D_{2/3}$Rs. A previous PET study in anesthetized cynomolgus monkeys using $[^{11}C]$ MNPA and $[^{11}C]$raclopride reported that the K_i values calculated on the basis of $D_{2/3}$R occupancy by apomorphine were 31 ng/ml for $[^{11}C]$MNPA and 29 ng/ml for $[^{11}C]$raclopride. These findings suggested that all $D_{2/3}$Rs are $D_{2/3}^{high}$ in vivo [33].

In human PET studies, $D_{2/3}^{high}$ binding is typically estimated with the subject in a conscious state; therefore, the dopamine hypersensitivity thought to be observed in patients with schizophrenia may not be detected with $D_{2/3}$R agonist ligands due to the low ratio of $D_{2/3}^{high}$ to $D_{2/3}^{low}$.

3.1.3 Dual-Ligand Approach for Estimating the Ratio of $D_{2/3}^{high}$ to $D_{2/3}^{low}$

The findings of several previous studies have suggested that the ratio of $D_{2/3}^{high}$ to $D_{2/3}^{low}$ can vary under pharmacological and physiological conditions. With respect to the dopaminergic system, chronic administration of antipsychotics (haloperidol, risperidone, olanzapine, quetiapine, and clozapine), catechol-O-methyltransferase (COMT) deficiency, dopamine D_4 receptor absence, sensitization by amphetamine, and trace amine-1 receptor activation can increase the proportion of $D_{2/3}^{high}$ two to four fold. Increases in the proportion of $D_{2/3}^{high}$ have also been reported due to factors not directly related to the dopaminergic system, such as cannabinoid receptor deficiency and ethanol withdrawal [34, 35].

Given that the ratio of $D_{2/3}^{high}$ to $D_{2/3}^{low}$ fluctuates in various physiological states, a dual-radioligand approach using $D_{2/3}R$ agonist radioligands and a $D_{2/3}R$ antagonist radioligand to estimate the ratio of $D_{2/3}^{high}$ to $D_{2/3}^{low}$ is needed to detect alterations in $D_{2/3}^{high}$ in schizophrenia patients. A previous study validating the dual-ligand approach with [11C]MNPA and [11C]raclopride showed that the binding capacity ratios of [11C]MNPA to [11C] raclopride in the human caudate nucleus and putamen were 0.24 ± 0.04 and 0.27 ± 0.03, respectively, with intrasubject variability of 0.79 and 0.8, indicating high reproducibility [24]. A dual-ligand approach using different doses of risperidone in healthy subjects showed no change in $D_{2/3}R$ occupancy [36].

3.2 Affinity States of $D_{2/3}Rs$ in the Pathophysiology of Schizophrenia

Both single- and dual-ligand approaches have been used in previous PET studies investigating the pathophysiology of $D_{2/3}^{high}$ in patients with schizophrenia. A single-ligand study using [11C] NPA in unmedicated patients with schizophrenia showed a slight increase in [11C]NPA binding compared to that in healthy controls [37]. This result is consistent with those of previous studies using $D_{2/3}R$ antagonist ligands to measure both $D_{2/3}^{high}$ and $D_{2/3}^{low}$ binding. However, PET studies using [11C]-(+)-PHNO in unmedicated schizophrenia patients did not reveal differences in binding compared to healthy controls [38]. In addition, no differences in [11C]-(+)-PHNO binding were found among patients with schizophrenia, patients with a clinically high risk for schizophrenia, and healthy controls during cognitive task performance [39]. On the other hand, significant differences in [11C]-(+)-PHNO binding have been found among these groups during a stress task [40, 41]. PHNO has a 25–48-fold greater affinity for dopamine D_3Rs than for D_2Rs, which may account for the observed differences [42].

In dual-ligand PET studies with [11C]MNPA and [11C]raclopride in unmedicated and drug-naive patients with schizophrenia, no significant differences were found in the binding potential of [11C]MNPA and [11C]raclopride compared to healthy controls. However, the ratio of [11C]MNPA to [11C]raclopride binding in the putamen was significantly higher, suggesting an increase in the ratio of $D_{2/3}^{high}$ to $D_{2/3}^{low}$ in the putamen. Furthermore, the binding potential of [11C]MNPA and [11C]raclopride in the caudate nucleus showed a significant positive correlation with the depression factor of the Positive and Negative Syndrome Scale [43].

3.3 Summary

A dual-ligand approach to estimate the ratio of $D_{2/3}^{high}$ to $D_{2/3}^{low}$ can be an alternative method to elucidate the pathophysiology of schizophrenia. From the perspective of the pathophysiology of schizophrenia, the elevation of synaptic dopamine levels may result in a greater occupancy of endogenous dopamine at $D_{2/3}^{high}$ than in healthy subjects. Meta-analyses have shown that drug-naive and unmedicated patients with schizophrenia exhibit increased presynaptic dopamine synthesis capacity [3]. Furthermore, $[^{11}C]$raclopride binding in the striatum after the depletion of synaptic dopamine with α-methylparatyrosine was significantly greater in patients with schizophrenia than in healthy controls [44], indicating the presence of elevated levels of endogenous dopamine in the synaptic cleft. An increased ratio of $D_{2/3}^{high}$ to $D_{2/3}^{low}$ in the striatum may be an additional cause of $D_{2/3}R$ hypersensitivity in patients with schizophrenia.

4   Notes

None

Acknowledgments

None

References

1. Seeman P et al (1987) Human brain D1 and D2 dopamine receptors in schizophrenia, Alzheimer's, Parkinson's, and Huntington's diseases. Neuropsychopharmacol Off Publ Am Coll 1:5–15

2. Laruelle M (1998) Imaging dopamine transmission in schizophrenia. A review and meta-analysis. Q J Nucl Med Off Publ Ital Assoc Nucl Med AIMN Int Assoc Radiopharmacol IAR 42:211–221

3. Howes OD et al (2012) The Nature of dopamine dysfunction in schizophrenia and what this means for treatment: meta-analysis of imaging studies. Arch Gen Psychiatry 69:776

4. Lieberman JA, Kane JM, Alvir J (1987) Provocative tests with psychostimulant drugs in schizophrenia. Psychopharmacology 91:415–433

5. Curran C, Byrappa N, McBride A (2004) Stimulant psychosis: systematic review. Br J Psychiatry J Ment Sci 185:196–204

6. Nordström AL, Farde L, Eriksson L, Halldin C (1995) No elevated D2 dopamine receptors in neuroleptic-naive schizophrenic patients revealed by positron emission tomography and [11C]N-methylspiperone. Psychiatry Res 61:67–83

7. Antonini A et al (1993) Effect of age on D2 dopamine receptors in normal human brain measured by positron emission tomography and 11C-raclopride. Arch Neurol 50:474–480

8. Rinne JO et al (1993) Decrease in human striatal dopamine D2 receptor density with age: a PET study with [11C]raclopride. J Cereb Blood Flow Metab Off J Int Soc Cereb Blood Flow Metab 13:310–314

9. De Lean A, Kilpatrick BF, Caron MG (1982) Dopamine receptor of the porcine anterior pituitary gland. Evidence for two affinity states discriminated by both agonists and antagonists. Mol Pharmacol 22:290–297

10. George SR et al (1985) The functional state of the dopamine receptor in the anterior pituitary is in the high affinity form. Endocrinology 117:690–697

11. Richfield EK, Penney JB, Young AB (1989) Anatomical and affinity state comparisons between dopamine D1 and D2 receptors in the rat central nervous system. Neuroscience 30:767–777

12. Sibley DR, De Lean A, Creese I (1982) Anterior pituitary dopamine receptors. Demonstration of interconvertible high and low affinity states of the D-2 dopamine receptor. J Biol Chem 257:6351–6361

13. Hall H, Sällemark M (1987) Effects of chronic neuroleptic treatment on agonist affinity states of the dopamine-D2 receptor in the rat brain. Pharmacol Toxicol 60:359–363

14. Gainetdinov RR et al (2003) Dopaminergic supersensitivity in G protein-coupled receptor kinase 6-deficient mice. Neuron 38:291–303

15. Pierre PJ, Vezina P (1998) D1 dopamine receptor blockade prevents the facilitation of amphetamine self-administration induced by prior exposure to the drug. Psychopharmacology 138:159–166

16. Seeman P, Tallerico T, Ko F, Tenn C, Kapur S (2002) Amphetamine-sensitized animals show a marked increase in dopamine D2 high receptors occupied by endogenous dopamine, even in the absence of acute challenges. Synapse 46:235–239

17. Seeman P (2005) An update of fast-off dopamine D2 atypical antipsychotics. Am J Psychiatry 162:1984–1985

18. Neumeyer JL, Gao Y, Kula NS, Baldessarini RJ (1990) Synthesis and dopamine receptor affinity of (R)-(−)-2-fluoro-N-n-propylnorapomorphine: a highly potent and selective dopamine D2 agonist. J Med Chem 33:3122–3124

19. Baldessarini RJ, Kula NS, Gao Y, Campbell A, Neumeyer JL (1991) R(-)2-fluoro-N-n-propylnorapomorphine: a very potent and D2-selective dopamine agonist. Neuropharmacology 30:97–99

20. Lahti RA et al (1996) Affinities and intrinsic activities of dopamine receptor agonists for the hD21 and hD4.4 receptors. Eur J Pharmacol 301:R11–R13

21. Gardner B, Strange PG (1998) Agonist action at D2(long) dopamine receptors: ligand binding and functional assays. Br J Pharmacol 124:978–984

22. Hwang DR, Kegeles LS, Laruelle M (2000) (-)-N-[(11)C]propyl-norapomorphine: a positron-labeled dopamine agonist for PET imaging of D(2) receptors. Nucl Med Biol 27:533–539

23. Finnema SJ et al (2005) A preliminary PET evaluation of the new dopamine D2 receptor agonist [11C]MNPA in cynomolgus monkey. Nucl Med Biol 32:353–360

24. Kodaka F et al (2013) Test-retest reproducibility of dopamine D2/3 receptor binding in human brain measured by PET with [11C] MNPA and [11C]raclopride. Eur J Nucl Med Mol Imaging 40:574–579

25. Wilson AA et al (2005) Radiosynthesis and evaluation of [11C]-(+)-4-propyl-3,4,4a,5,6,10b-hexahydro-2H-naphtho [1,2-b][1,4]oxazin-9-ol as a potential radiotracer for in vivo imaging of the dopamine D2 high-affinity state with positron emission tomography. J Med Chem 48:4153–4160

26. Graff-Guerrero A et al (2008) Brain region binding of the D2/3 agonist [11C]-(+)-PHNO and the D2/3 antagonist [11C]raclopride in healthy humans. Hum Brain Mapp 29:400–410

27. Subburaju S, Sromek AW, Seeman P, Neumeyer JL (2018) New dopamine D2 receptor agonist, [3H]MCL-536, for detecting dopamine D2high receptors in vivo. ACS Chem Neurosci 9:1283–1289

28. Narendran R et al (2005) Measurement of the proportion of D2 receptors configured in state of high affinity for agonists in vivo: a positron emission tomography study using [11C]N-propyl-norapomorphine and [11C]raclopride in baboons. J Pharmacol Exp Ther 315:80–90

29. Narendran R et al (2004) In vivo vulnerability to competition by endogenous dopamine: comparison of the D2 receptor agonist radiotracer (-)-N-[11C]propyl-norapomorphine ([11C]NPA) with the D2 receptor antagonist radiotracer [11C]-raclopride. Synapse 52:188–208

30. Ginovart N et al (2006) Binding characteristics and sensitivity to endogenous dopamine of [11C]-(+)-PHNO, a new agonist radiotracer for imaging the high-affinity state of D2 receptors in vivo using positron emission tomography. J Neurochem 97:1089–1103

31. Seneca N et al (2006) Effect of amphetamine on dopamine D2 receptor binding in nonhuman primate brain: A comparison of the agonist radioligand [11C]MNPA and antagonist [11C]raclopride. Synapse 59:260–269

32. Finnema SJ et al (2014) (18)F-MCL-524, an (18)F-Labeled Dopamine D2 and D3 Receptor Agonist Sensitive to Dopamine: A Preliminary PET Study. J Nucl Med Off Publ Soc Nucl Med 55:1164–1170

33. Finnema SJ et al (2009) Dopamine D(2/3) receptor occupancy of apomorphine in the nonhuman primate brain--a comparative PET

study with [11C]raclopride and [11C]MNPA. Synap N Y N 63:378–389

34. Seeman P et al (2006) Psychosis pathways converge via D2High dopamine receptors. Synapse 60:319–346

35. Seeman P et al (2005) Dopamine supersensitivity correlates with D2High states, implying many paths to psychosis. Proc Natl Acad Sci USA 102:3513–3518

36. Kodaka F et al (2011) Effect of risperidone on high-affinity state of dopamine D2 receptors: a PET study with agonist ligand [11C](R)-2-CH3O-N-n-propylnorapomorphine. Int J Neuropsychopharmacol 14:83–89

37. Frankle WG et al (2018) Amphetamine-Induced Striatal Dopamine Release Measured With an Agonist Radiotracer in Schizophrenia. Biol Psychiatry 83:707–714

38. Graff-Guerrero A et al (2009) The Dopamine D2 Receptors in High-Affinity State and D3 Receptors in Schizophrenia: A Clinical [11C]-(+)-PHNO PET Study. Neuropsychopharmacology 34:1078–1086

39. Suridjan I et al (2013) Dopamine D2 and D3 binding in people at clinical high risk for schizophrenia, antipsychotic-naive patients and healthy controls while performing a cognitive task. J Psychiatry Neurosci JPN 38:98–106

40. Mizrahi R et al (2012) Increased stress-induced dopamine release in psychosis. Biol Psychiatry 71:561–567

41. Setiawan E et al (2018) Association of translocator protein total distribution volume with duration of untreated major depressive disorder: a cross-sectional study. Lancet Psychiatry 5:339–347

42. Gallezot J-D et al (2012) Affinity and selectivity of $[^{11}C]$-(+)-PHNO for the D3 and D2 receptors in the rhesus monkey brain in vivo. Synap. N. Y. N 66:489–500

43. Kubota M et al (2017) Affinity states of striatal dopamine D2 receptors in antipsychotic-free patients with schizophrenia. Int J Neuropsychopharmacol 20:928–935

44. Kegeles LS et al (2010) Increased synaptic dopamine function in associative regions of the striatum in schizophrenia. Arch Gen Psychiatry 67:231–239

Chapter 9

Finding the Fire: Mapping Brain Temperature Elevations as a Surrogate of Focal Neuroinflammation

Ayushe A. Sharma, Rodolphe Nenert, and Jerzy P. Szaflarski

Abstract

A persistent, low-level neuroinflammatory state is a pathophysiological characteristic of multiple sclerosis, Alzheimer's disease, and epilepsy, among other neurological disorders. Elucidating the mechanisms underlying neuroinflammation is crucial for understanding the pathophysiology of these disorders, monitoring their progression, and developing effective treatments. It is equally important to find a way to visualize neuroinflammation in vivo, particularly given the scarcity of reliable techniques. Focal, sustained brain temperature elevations are directly related to various peripheral cellular and molecular biomarkers of inflammation, and serve as an indirect proxy of increased metabolic activity associated with neuroinflammation. Brain temperature elevations can be noninvasively visualized using volumetric magnetic resonance spectroscopic imaging and thermometry (MRSI-t) based on the chemical shift of water. In addition to reviewing the MRSI-t analytical framework and methodology, this chapter addresses applications of MRSI-t, including how it can be used to localize brain temperature elevations indicative of central nervous system damage and track treatment response.

Key words Neuroinflammation, Brain temperature mapping, Magnetic resonance spectroscopic imaging and thermometry, Seizure disorders

---

1 Introduction

1.1 Neuroinflammation: A Circuitous Process that Plays a Key Pathophysiological Role in Many Neurological Disorders

Chronic neuroinflammation (NI) is a significant pathological factor in the development and progression of several neurological disorders, including multiple sclerosis (MS), Alzheimer's disease (AD), traumatic brain injury (TBI), and epilepsy [1–4]. The pathophysiology that triggers these disorders is diverse, ranging from autoimmune processes to tissue insult/injury, and remains poorly understood. One key factor unites these disorders despite their diverse etiologies: an anomaly in the defense mechanisms that *repair* brain damage in order to *prevent* brain dysfunction [5].

Neurons/neural tissues respond to a primary mechanical insult (e.g., TBI infection, low oxygen levels) or secondary injury by releasing damage-associated molecular patterns (DAMPS) into

Daichi Sone (ed.), *Molecular Imaging for Brain Diseases*, Neuromethods, vol. 222, https://doi.org/10.1007/978-1-0716-4494-2_9,
© The Author(s), under exclusive license to Springer Science+Business Media, LLC, part of Springer Nature 2025

the extracellular space [6]. DAMPS, along with other factors, eventually trigger an inflammatory response [6]. Adenosine triphosphate (ATP), for example, activates dormant microglia to migrate to the injury site [7]. Upon arrival, these activated microglia release various pro-inflammatory cytokines, such as nitrous oxide, interleukin-6 (IL-6), tumor necrosis factor alpha (TNF-α), and interleukin-1-beta (IL-1β) [8–11].[1] These molecules collectively trigger a cascade of connected and synchronized reactions, drawing additional cells to drive the tissue repair process and, if necessary, directly attack the source of tissue damage (e.g., a virus or bacterium). TNF-α, for example, plays a major role in this context by repairing damaged myelin [12]. Acute tissue damage-induced NI is beneficial and neuroprotective, but chronic NI that persists unnecessarily is harmful and neurotoxic. In MS, the immune system mistakenly perceives myelin as a threat and attacks it, leading to the destruction of healthy myelin [13, 14]. In post-traumatic focal epilepsy, acute NI triggered by tissue damage likely becomes chronic [15–17]. Regardless of the cause, chronic NI impairs the function of the blood–brain barrier, leads to neuronal death, and ultimately results in a vicious cycle of neurodegeneration [5, 6, 9, 18–21].

Sustained activation of microglia can induce structural, functional, and biochemical neurodegeneration by triggering the chronic activation of other neuroinflammatory cells [5, 22, 23]. While microglia sustain the inflammatory response, it is important to acknowledge that chronic NI is not *solely* caused by these cells [24, 25]. Other contributors include, e.g., the release of pro-inflammatory cytokines, particularly TNF-α, that can exhibit both, restorative and degenerative properties [24, 25]. Identifying what propels and maintains microglial activity and the release of these pro-inflammatory cytokines is a difficult yet crucial objective in the field. Understanding the drivers and sustainers of NI is paramount, especially in disorders where chronic, low-level NI yields detrimental outcomes. However, the complexity of NI, which encompasses a variety of overlapping and intersecting events, makes it challenging to pinpoint the exact "how" and "why" of the problem. While understanding these aspects is important, questions about "when," "where," and "to what extent" might have a more immediate impact on patient outcomes. A neuroimaging-based biomarker of NI could potentially address these questions, allowing for the assessment of disease progression, treatment response, and many associated processes, such as cognitive impairment or comorbid mood issues [22, 23, 26–31].

---

[1] Note: This is meant to be an overview; the process is, of course, much more intricate and nuanced than can be summarized here.

1.2 The Search for a Noninvasive, Imaging-Based Biomarker of Neuroinflammation

Despite the prevalence of NI in neurological disorders, its in vivo visualization remains challenging. NI is, by its very nature, a circuitous, integrated, interconnected, and multifaceted process with both spatial and temporal variations [5]. Although NI typically starts as a response to tissue damage in isolated or widespread cortical areas, the transition from an acute neuroprotective state to a chronic neurotoxic state is not fully understood [5, 13]. Isolating a key process or cell is difficult due to the array of diverse and interconnected cellular actors involved in neuroinflammatory processes, as well as their temporal and spatial variations [5]. Furthermore, there are several layers of "what" within a particular cell type, so it is not just a matter of "which," "when," "how," and "where" cells are activated [5]. For instance, microglia are arguably straightforward mechanistic targets because they serve as both the conductor *and* metronome for orchestrating NI. But there are several types of microglia. Microglia can be categorized based on whether they are dormant or active, yet this only scratches the surface of their functional and phenotypic heterogeneity. The cellular composition of their environment determines whether microglia are polarized to serve a neuroprotective or neurodegenerative phenotype [6, 32]. These conceptual difficulties are linked to methodological and technological shortcomings. While methods for localizing NI are well established in animal studies, translating these methods to humans is challenging due to the need for direct access to brain tissue, which is uncommon in vivo for disorders other than focal epilepsy or brain tumors. Animal studies of NI also necessitate a NI trigger. While these studies simulate the consequences of NI, they cannot fully model how, when, where, and why each patient develops chronic NI *when the pathophysiological cause of the disorder remains unknown.*

There are ways to measure NI, but they are not without limitations. For example, cerebrospinal fluid (CSF) analysis can be used to measure the concentration of pro-inflammatory cells and molecules present in the fluid bathing the brain. However, the method is less than ideal given the lack of spatial information and the highly invasive lumbar puncture needed to obtain CSF. While decades of NI-induced neuronal damage may yield macroscopic structural abnormalities, NI-induced changes in tissue microstructure or biochemical phenomena are not detectable by standard structural MRI. Positron emission tomography (PET) is the most thoroughly investigated imaging technique for visualizing NI, and it is arguably the "gold standard" among the currently validated imaging tools [33–43]. The most unique characteristic of PET is that it visualizes the binding of a radiotracer to a target receptor of interest. The idea that 18-kDa translocator protein (TSPO) was upregulated on active microglia led to the development of approaches that used TSPO-PET to visualize focal NI [44]. However, this method faced challenges due to (1) the inability to isolate a specific microglial subtype, (2) the discovery that TSPO is expressed on microglia and other cells at various stages in the neuroinflammatory process and throughout disease progression, (3) variable binding among individuals, and (4) bioavailability issues and challenges in absolute quantification [10, 45–47]. Therefore, while TSPO-PET can provide a generalized view of NI, its high cost and methodological challenges limit its widespread clinical application [37, 43, 47–50].

1.3 Brain Temperature Elevations: A Marker of Neuroinflammation

Due to the challenges in identifying neurotoxic and active cells specific to NI, brain temperature serves as a practical representative of the collective metabolic activity of cells active in neuroinflammatory processes [51–53]. The aggregate effect of these cells is more potent than that of individual parts, which justifies focusing on a macroscopic by-product such as abnormal brain temperature increases. Therefore, brain temperature mapping could be an effective tool for diagnosing and monitoring neurological disorders characterized by metabolic and homeostatic disruptions, as the overall activity of neuroinflammatory cells in an NI-affected cortical area can cause local temperature changes [52, 54].

Despite extensive literature on various aspects of the brain's structure and functions, our understanding of brain temperature remains limited. Brain temperature fluctuations are a physiological reality influenced by metabolic neural activity, yet they are quite distinct from traditional measures of neural activity. Aristotle first described the brain's temperature regulation characteristics, although it was in the context of how he believed the brain cools the heart [51, 55]. The first thermal recordings from an awake animal's brain were conducted in the late 1860s, which intriguingly predates the first recordings of the brain's electrical activity [53]. Early cerebral thermometry studies demonstrated that brain temperature can vary significantly due to environmental challenges, spontaneous changes in activity states, and different behaviors, and is highly dependent on an animal's motivational state [53, 56 59]. While the precise boundary between normal and abnormal brain temperature is still not entirely clear, studies have generally established ~37–38 °C as mean global brain temperature in healthy adults, with 40–42 °C considered definitely abnormal [60–64]. In the healthy state, brain temperature typically exceeds systemic temperature by 0–1 °C, and a low–high gradient is observed from deeper subcortical structures to more superficial cortical tissue [52, 61, 65]. Brain activity also has the potential to induce local heat production [52, 54, 61]. Neuronal activation, hormonal changes in various contexts (e.g., menstrual cycles), vaccinations, and inflammation—both peripheral and cerebral—can cause transient physiological fluctuations in brain temperature [52, 54, 61]. However, given the brain's enclosed environment and the primary role of cerebral blood flow in heat dissipation, long-term changes in brain temperature can occur due to excessive heat production and/or inadequate brain cooling [65]. The sharp rise in metabolic activity caused by pathological states, such as chronic, low-level NI, characterized by a cascade of looping, seemingly never-ending cellular events, can overwhelm the brain's already limited cooling mechanisms and raise brain temperature an additional 1–2 °C above core body temperature [65–68]. Brain temperature rises coincide with neuroinflammatory phenomena such as leukocyte extravasation and accumulation, blood–brain barrier permeability, and cerebral edema [69–75]. Just as high brain temperature (i.e., hyperthermia) is harmful, hypothermia may be therapeutic and has been shown to attenuate NI-related pathologies immediately after tissue injury caused by stroke and cardiac arrest [51, 69, 76–78].

2 Brain Temperature Mapping: Analytical Framework and Key Methodological Considerations

2.1 Volumetric Magnetic Resonance Spectroscopic Imaging and Thermometry (MRSI-t)

Volumetric magnetic resonance spectroscopic imaging (MRSI) and thermometry (MRSI-t) is the most noninvasive way to indirectly image NI, with proven reliability and reproducibility for repeated scans spaced up to 12 weeks apart [60, 62, 63, 79–81]. Compared to single-voxel MRSI, which can only calculate metabolite levels in a predetermined region of interest, volumetric or whole-brain MRSI represents a significant advancement in spectroscopic imaging [23]. Volumetric MRSI allows measuring metabolite levels and characterizing brain temperature throughout the brain, based on data from echo-planar sequence imaging (EPSI) [63, 81]. This advancement is especially important for imaging disorders where a priori selection of a brain region is not possible, especially since not all neurological disorders are caused by pathologies in a single brain region. Although MRSI has a longer scan time and lower resolution than single-voxel techniques, it compensates for these drawbacks by providing a richer, more comprehensive picture of brain metabolism and homeostasis. The correlation between MRSI-derived brain temperature and implanted probe recordings underscores its clinical potential in assessing NI-related temperature changes and guiding therapeutic interventions [82].

The methods for processing and visualizing MRSI-t-based T_{CRE} data have been previously described in detail [60, 83, 84]. However, due to word limits imposed by scientific publications, there is less information available regarding methodological considerations and other critical steps necessary for successful data collection for MRSI-t analyses. To that end, Subheading 2.2 provides a detailed overview of the steps that must be taken to ensure high-quality data collection, followed by a brief summary of MRSI-t data processing and analysis approaches in Subheading 2.3.

**2.2   Acquiring High-
Quality MRSI-t Data:
Methodological
Considerations and
Recommendations**

In order to ensure high-quality data, there are several methodological considerations that need to be taken into account.

*2.2.1   Planning an
Effective Sequence of
Scans*

The order of scans is critical to ensure data are collected as efficiently as possible. In our studies, we apply the following sequence order on a Siemens Magnetom 3 T Prisma using a 20-channel head coil:

1. *Localizer*: The localizer scan (<1 min) orients the brain's position in subsequent scans, ensuring they are all aligned in the same plane.

2. *Structural MRI (e.g., T1-MPRAGE)*: The structural scan (6–8 min) is used to register processed MRSI-t data to an atlas template to identify specific regions of interest.

3. *Resting-state fMRI (rs-fMRI)*: This scan captures the brain's activity when it is not engaged in a specific task. We included a ~ 6-min resting-state fMRI before the EPSI scan to allow time for reslicing the patient's T1-weighted image and positioning the saturation band before commencing EPSI, both of which will be discussed in the sections that follow.

4. *Echo-planar sequence imaging (EPSI)*: After shimming is completed (~2–4 min), this ~17-min scan gathers water-suppressed metabolite data and reference water signals. The collected data are reconstructed, integrated, and processed by MRSI-t to visualize both global and regional metabolite concentrations, as well as T_{CRE}.

Due to the length of the scans, it is important that participants are comfortable. Because we do not collect task-related fMRI for our MRSI-t studies, we encourage participants to select a TV show or film to watch in the scanner. This not only increases participant comfort, but also significantly reduces the occurrence of motion artifacts. This is critical as even small motion artifacts can affect both shimming and EPSI data quality.

*2.2.2   Understanding and
Manipulating the
Saturation Band*

During EPSI data acquisition, signals from water-containing structures in the head are typically masked with an oblique saturation band. Unfortunately, due to the lack of sophisticated and more selective technology, the saturation band is a three-dimensional (3D) rectangular slab with a fixed shape and size. The only adjustments that can be made are to the position and angle. The angling of the saturation band is used to cover the eyes and sphenoidal sinuses, thereby blocking the signal from these water-containing structures. Since it cannot be reshaped, and since each participant has different structural anatomy, the saturation band may mask signal from frontal cortical areas. Additionally, the placement and orientation of the saturation band are completed for every scan, and the parameters cannot typically be saved. Therefore, its position cannot be consistently replicated across multiple scans. These limitations are important considerations, as regions where the saturation band cuts off signal result in missing or contaminated metabolite and temperature data for impacted regions of interest.

Summary of Steps for Effective Saturation Band Placement

- Change the angle to be as steep as possible.

- Position the band to completely exclude the aqueous humor of both eyes and the sphenoidal sinuses, but retain as much gray matter as possible.

- Use the resliced axial, coronal, and sagittal T1 images (*see* Subheading 2.2.3) to navigate through slices and determine if the band needs adjustment in location or angle.

2.2.3 Reslicing for Better Visualization of Saturation Band Placement

To facilitate the placement of the saturation band, we digitally reslice the coronal, axial, and sagittal views of the T1 to aid in seeing higher quality images in brain regions where the saturation band will be placed.

- One recommended adjustment would be to keep the image thickness at 0.8, but adjust the bounding box and set the distance between images to 1.0 mM. We also recommend increasing the number of images either manually or by altering the bounds of the box covering the range of axial slices where the saturation band will be positioned over the eyes.

- This adjustment aims to increase the number of slices while decreasing the thickness of slices in the very regions where the saturation band will be placed. Once these modified axial, coronal, and sagittal slices are saved, they can be used to help visualize optimum saturation band placement.

2.2.4 Automatic and Interactive Shimming Optimize MRSI-t Data Quality

Shimming during EPSI acquisition can substantially reduce magnetic field inhomogeneities and improve the signal-to-noise ratio by adjusting the spectral linewidth. While the process is time-consuming and requires substantial training, it can greatly improve data quality. The goal is to achieve a smaller full-width half-maximum (FWHM), with an ideal target of 20–25. To achieve this goal, we recommend performing shimming with a two-step process:

1. *Automated 3D shimming*: The first step is an automated procedure run by the scanner control computer that adjusts the magnetic field to minimize distortions.

2. *Interactive shimming*: Following the automated process, we highly recommend proceeding with interactive shimming to further optimize signal resolution and data quality. This step requires the operators to complete extensive training in manual shimming. Although 3D-automatic shimming is typically where most investigators stop, manual shimming allows further fine-tuning the FWHM to significantly optimize data quality. However, completing the training required for manual shimming is a challenge.

Here is an overview of the step-by-step process for shimming:

1. *Place the saturation band*: This should be done immediately after the rs-fMRI scan, but before running the EPSI sequence, following the methods in Subheadings 2.2.2 and 2.2.3.

2. *Automated shimming*: Once the saturation band is in place, start automated shimming.

3. *Manual shimming*: Manually adjust all gradients (X, Y, and Z) one at a time. For large adjustments, start with the Z gradient; for smaller increments, start with the X gradient. With every set of three to four adjustments, wait for the FWHM to level out and evaluate whether the FWHM is improving.

4. *Load the best shim*: After each adjustment, click "Load Best" shim and move on to the next gradient. Stop when additional adjustments to the X, Y, and Z gradients no longer improve the FWHM.

5. *Record the final FWHM value*: We recommend keeping a written record of the FWHM value achieved by each shim on the study protocol sheet. This will come in handy for data quality checks.

6. *Start the EPSI scan*: Close the shimming window, and start the EPSI scan.

The goal is to achieve a smaller FWHM, ideally 20–25. However, since this is not always attainable, the maximum peak FWHM of all participants' data should not exceed 30 Hz [85]. Shimming occurs during the break between two loud MR sequences and produces little noise. Unfortunately, these breaks are precisely when participants typically move, stretch, etc. Before performing shimming, we ask participants to take a brief rest and stretch, and proceed only after confirming that they are ready. Before we begin shimming, we participants that data are still being collected even if they do not hear the scanner. For example, we typically say the following:

"We are moving on to the next phase of our study, which includes the longer 17-minute scan. I'll be turning your [movie/tv show] back on. The first few minutes are without the loud noises you've been hearing up until this point. Please stay as still as possible from now until the end of the 17-minute scan. Even if you don't hear the scanner making loud noise, we are still collecting data."

2.3 Approaches for MRSI-t Post-processing, Statistical Analyses, and Visualization Depend on Your Research Aims

There are various methods for acquiring and analyzing whole-brain EPSI data, ranging from the choice of sequence and scanner to the software tool. Selecting the right software for processing and analyzing MRSI data is critical. The decision largely depends on your specific project goals, budget constraints, and level of computing experience. Key spectral fitting and processing software tools for quantifying metabolite concentrations include:

2.3.1 Software Considerations

1. LCModel (Linear Combination of Model Spectra).
2. SIVIC (Spectroscopic Imaging, Visualization, and Computing).
3. Tarquin.
4. jMRUI (Java-based Magnetic Resonance User Interface).
5. MIDAS (Magnetic Resonance Imaging and Data Analysis System).

In our lab, we use MIDAS, an open-source tool for comprehensive processing, analysis, and visualization of MRSI data developed by Maudsley and colleagues [63, 81]. MIDAS provides a user-friendly interface, with dedicated module for voxel-wise brain temperature estimation and easy data export for more advanced analyses using other tools [63, 81]. Based on this, MIDAS is a great tool for researchers primarily interested in generating brain temperature maps and conducting more detailed analyses of temperature fluctuations across different brain regions.

2.3.2 Estimating and Extracting Brain Temperature Data

As previously noted, the methods for acquiring, analyzing, and preprocessing data have been described [60, 63, 80, 83, 84, 86, 87]. Therefore, we recommend reviewing the previously published processing, post-processing, and visualization techniques or visiting our lab wiki (https://tinyurl.com/mwcnestm). Regardless of the tool used, validated pipelines are well-documented, and each of these methods are conceptually linked by how brain temperature is quantified. For any sequence or processing software you use, noninvasive MRSI-based brain temperature measurements are based on the chemical shift of the temperature-sensitive peak of water, a shift that is best quantified by comparison to the temperature-insensitive peak of a reference metabolite. Reference metabolites such as N-acetylaspartate (NAA), creatine phosphate (CRE), and the combined signal from choline-containing compounds (CHO) consistently appear at the same location on the

MRSI-t data processing: overview

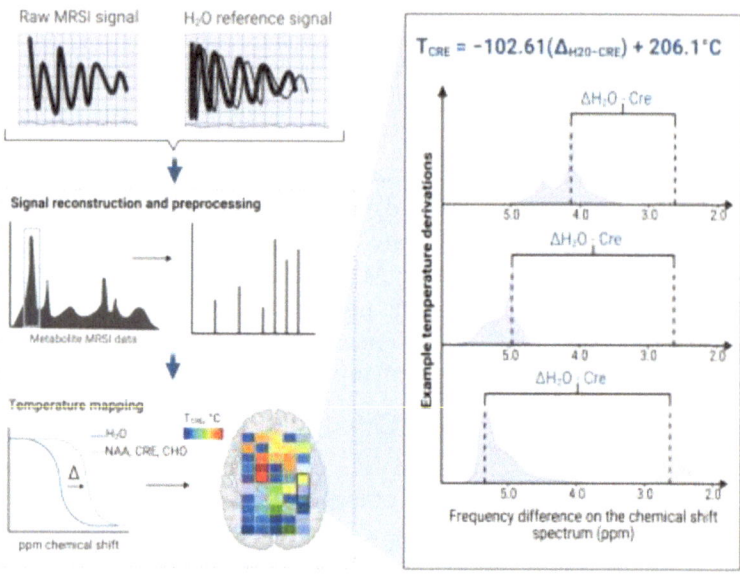

Fig. 9.1 Overview of processing and brain temperature estimation using volumetric magnetic resonance spectroscopic imaging and thermometry (MRSI-t). Raw metabolite data provide spectroscopic information necessary for locating temperature-insensitive reference metabolite peaks, while the signal of water provides the location of the temperature-sensitive peak of water. N-acetylaspartate (NAA), creatine phosphate (CRE), and choline-containing compounds are typically used as reference metabolites. For our studies, we used CRE as the reference due to its stability across regions and both gray and white matter. The equation used to calculate brain temperature in each voxel, as well as example temperature-dependent shifts on the chemical shift spectrum, is shown on the left. This figure is a hybrid adaptation of two previously published figures: (1) Fig. 1 from Sharma et al. (2020) and (2) Fig. 1 from Sharma et al. (2023)

chemical shift spectrum, irrespective of brain temperature [63, 81]. The choice of reference metabolite should be based on the sample population (i.e., the disease in question) and which metabolite is most evenly distributed between brain regions and gray/white matter. However, the peak of water shifts its location on the ppm spectrum with changes in brain temperature, with a 0.01 ppm shift corresponding to a 1.0°C change [60, 62, 64, 82]. In our studies, we used CRE as the reference metabolite due to its abundance across brain regions and tissues [63]. Its stability on the ppm spectrum also leads to more reliable results on repeated [63]. Using the brain temperature estimation module in MIDAS, we estimated brain temperature in each voxel by calculating the frequency difference between the temperature-sensitive water peak and the temperature-independent peak of CRE with the following equation (Fig. 9.1): $T_{CRE} = -102.61(\Delta_{H20 - CRE}) + 206.1$ °C [63, 81, 88, 89].

After MRSI-t data processing is complete, T_{CRE} maps are extracted and further analyzed. It is recommended to impose quality measures when extracting data, e.g., excluding voxels if (1) they have a fitted metabolite linewidth >13 Hz (i.e., poor spectral quality), (2) they have outlying values >2.5 standard deviations away from voxels in the image, and (3) they contain less than 80% gray or white matter or > 30% CSF contribution to the voxel volume [90]. Once T_{CRE} are exported, each participant's data should be registered to their structural MRI data and then to a template atlas to allow pinpointing variations in T_{CRE}. In addition to the extraction of T_{CRE} maps, inverse spatial transformation can be applied to obtain atlas-defined ROIs in subject space. This step allows leveraging integrated spectra to obtain a single mean spectrum in the 47 ROIs outlined by the modified Automated Anatomical Labeling atlas, specifically developed for low-resolution MRSI data [91]. While the 3D T_{CRE} maps permit more compelling group analyses, the averaged spectra have a higher signal-to-noise ratio and are thus more accurate and less prone to outliers.

2.3.3 Post-processing and Statistical Analysis

Once T_{CRE} data are registered to structural and atlas templates, various post-processing techniques and statistical analyses can be employed to derive meaningful insights. Following data export, the choice of post-processing techniques depends on the specific research question, patient population, and objectives. While tools like MIDAS offer specialized functionality for brain temperature estimation, statistical and image calculation tools included in *FSL* (FMRIB Software Library), *SPM* (Statistical Parametric Mapping), and *AFNI* (Analysis of Functional NeuroImages) provide additional capabilities for advanced data analysis and visualization following data export.

Common approaches include the following:

- *Region of interest (ROI) analysis:* Extracting T_{CRE} values from predefined ROIs allows for the quantitative assessment of temperature variations within specific brain regions. For example, a left temporal lobe mask could be used to extract mean T_{CRE} for a group of participants; such a mask could also be used to restrict statistical analyses to this ROI.

- *Group comparisons:* Statistical tests, such as t-tests or ANOVA, can be used to compare T_{CRE} differences between groups (e.g., patients vs. controls) or conditions (e.g., pre- vs. post-treatment) using FSL, AFNI, or SPM. Additionally, more specialized SPM-based toolboxes can support multimodal (Fusion ICA toolbox developed by Vince Calhoun and colleagues) or multivariate analyses (Multivariate Repeated Measures [MRM] toolbox).

- *Correlation analysis:* Exploring relationships between voxelwise T_{CRE} and clinical variables, such as disease severity or cognitive performance, can provide valuable insights into disease mechanisms and prognosis.

Effective visualization of temperature data is crucial for interpreting results and communicating findings. Various techniques can be employed, including:

- *Heatmaps:* Superimposing T_{CRE} maps onto structural MRI images provides anatomical context and facilitates the identification of T_{CRE} abnormalities within specific brain structures with color-coded temperature variations across different brain regions. Heatmaps can also be used to specifically display T_{CRE} elevations above a certain threshold. For example, we set 38.5 ° C as the minimum threshold for visualizing atypical T_{CRE} elevations, because this corresponds to 2.5 standard deviations from the normative mean of 37.2 °C found in our previous studies [60]. Regardless of the approach, heatmaps are an excellent way to display both group trends and patient-specific data.

- *3D rendering:* Creating 3D renderings of T_{CRE} data enables comprehensive visualization and exploration of spatial patterns and uses similar methods to heatmap creation. However, the main difference is that they are not limited to a section of the brain, but instead show the brain as a 3D structure.

To ensure clinical relevance for group analyses, standardization of parameters—for example, lateralization in temporal lobe epilepsy (TLE)—is essential. When performing group analyzes in a MRSI-t study involving, for example, patients with TLE and healthy controls, it is important to ensure that the T_{CRE} data of TLE participants are flipped so that all participants have their ictal onset zone on the same side. For example, this is particularly useful if your TLE group consists of 20 participants, 15 of which have a left-sided onset. Flipping the data so that right-onset individuals (i.e., the smallest group) have a left-sided onset helps ensure that any results based on lateralization are not confounded. In cases where data flipping is necessary, it is advisable to also flip healthy controls' data in the same proportions to limit the introduction of left-right biases into analyses. Additionally, to limit the results, it may be optimal to perform this analysis with a left temporal lobe mask. Finally, for the consistency and alignment of multimodal imaging data that combines analysis of T_{CRE} maps with other imaging data, it is important to normalize T_{CRE} maps to other imaging data. This guarantees that the resolution or features of one modality do not obscure the results obtained from another.

3   Applications: Brain Temperature Elevations in Healthy and Diseased States

The studies conducted to date provide compelling evidence for the feasibility and potential clinical utility of MRSI-t for mapping brain temperature in healthy and diseased brains, particularly in the context of focal treatment-resistant epilepsy (TRE). While NI manifests itself differently in different disorders, epilepsy provides a unique model in which NI tends to be localized (*see* Box 1) and can be pinpointed as part of the surgical evaluation process, in contrast to the diffuse nature of NI in AD or MS.

One of the most significant findings in our MRSI-t studies to date was the localization of MRSI-t T_{CRE} elevations to seizure-producing regions in the mesial temporal lobe of patients with TLE (*see* Fig. 9.2a) [84]. Furthermore, focal T_{CRE} elevations

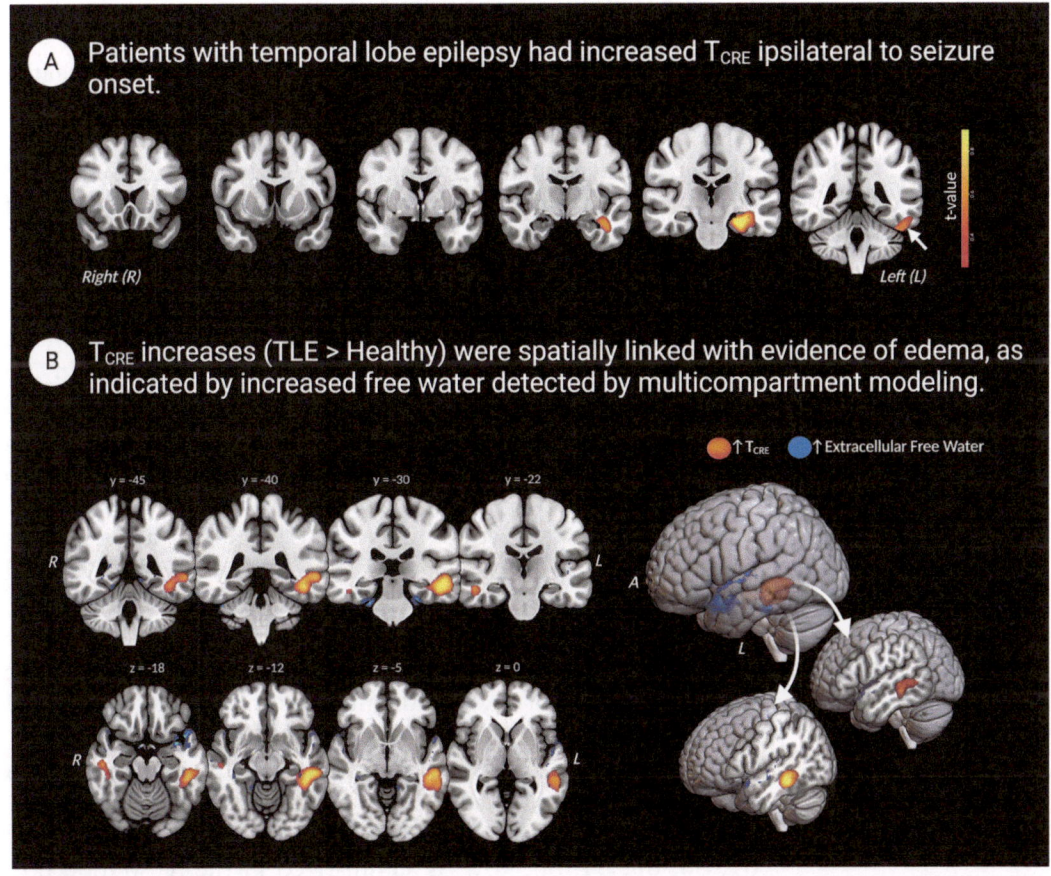

Fig. 9.2 Evidence of MRSI-t-based brain temperature elevations in seizure-producing regions when comparing patients with temporal lobe epilepsy to healthy control participants. Figure adapted using Biorender.com, with permission from *Epilepsia*. (Original citation: Sharma AA, Nenert R, Goodman A, Szaflarski JP. Concordance between focal brain temperature elevations and focal edema in temporal lobe epilepsy. Epilepsia. 2023 May;64(5):1289–1304. doi: 10.1111/epi.17538. Epub 2023 Mar 7. PMID: 36762949)

(38.5–42 °C) in the mesial temporal lobe (TLE > Healthy) were spatially concordant with evidence of edema in adjacent and/or nearby areas as indicated by neurite orientation density and dispersion imaging or NODDI (*see* Fig. 9.2b) [84]. The concordance between MRSI-t and NODDI findings was particularly noteworthy given that NODDI is a histopathologically verified tool, which further validates the use of MRSI-t as a method for noninvasively mapping pathophysiology [84, 92, 93]. These findings are also consistent with the literature showing that structural abnormalities in neurons and blood–brain barrier leakage occur in situations where both hyperthermia and edema are found in parallel [53]. The next phase in this series of studies examined whether T_{CRE} elevations in patients with TRE decreased following targeted anti-neuroinflammatory treatment with pharmaceutical-grade cannabidiol (CBD) [87]. Patients with TRE experienced a reduction in T_{CRE} within their ictal onset zones after 12 weeks of CBD, suggesting that CBD may be effective in reducing NI in these patients [87].

In another study, MRSI-t was used to map T_{CRE} abnormalities in patients with non-epileptic seizures, revealing preliminary evidence of more diffuse, widespread T_{CRE} abnormalities in regions implicated in dysfunctional emotion regulation, anxiety/fear, and sensory-motor integration [83]. A subsequent MRSI-t study of PNES and healthy controls found a complex pattern of T_{CRE} abnormalities associated with depression, quality of life, psychological symptoms, and disability. This included (1)↑T_{CRE} in the orbitofrontal cortex and angular cingulate gyrus alongside and (2)↓ T_{CRE} in the precentral gyrus, cerebellum, and areas within the occipital cortex [94]. Similarly, a MRSI-t study of healthy controls and patients with TLE both with and without depressive symptoms found elevated choline levels correlating with T_{CRE} abnormalities in the temporal (↑T_{CRE}) and frontal (↑T_{CRE} *and* ↓ T_{CRE}) lobes, supporting the intersection of NI and metabolic abnormalities shared in epilepsy and depression [86]. More recently, we used MRSI-t to map pre- and post-COVID changes in brain temperature. The most significant pre- to post-COVID T_{CRE} increase (mean 1.75 °C) was observed in the left olfactory tubercle in the primary olfactory cortex, with the most prominent increases found in those with chronic olfactory dysfunction [95].

Overall, MRSI-t-based increases in T_{CRE} have been found in various contexts, including focal epilepsy, chronic fatigue syndrome, neonatal encephalopathy, functional seizures, and more [61, 80, 83, 84, 86, 96]. As summarized in the illustration presented in Fig. 9.3, the MRSI-t studies conducted to date indicate that T_{CRE} increases are tied to other indicators of NI-induced damage. Based on these findings, MRSI-t could be a useful biomarker for noninvasively visualizing NI, a key factor in the development and progression of any disease with a neuroinflammatory component.

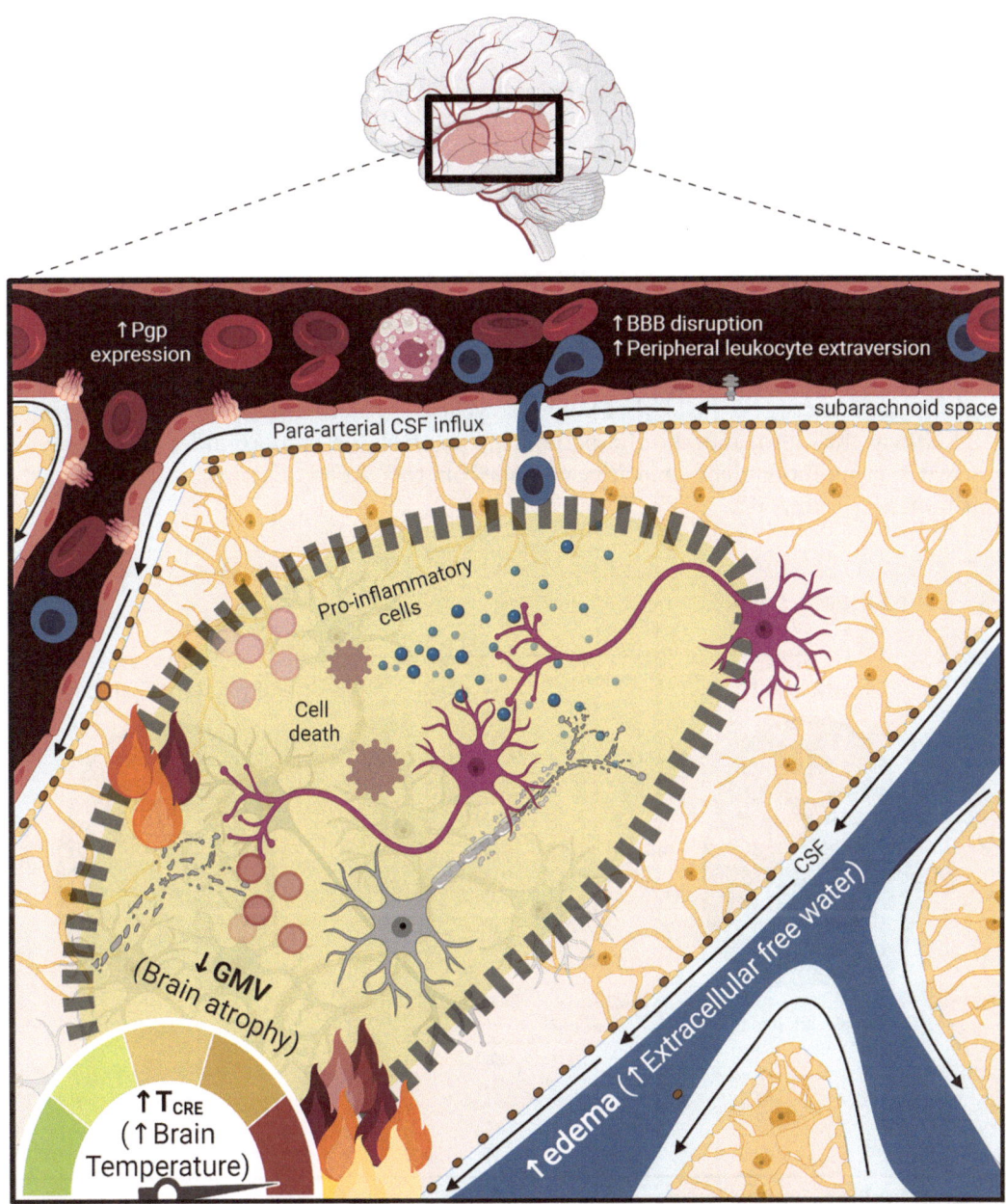

Fig. 9.3 MRSI-t allows mapping brain temperature elevations indicative of local NI. It can be used in various contexts where local damage stemming from chronic NI corresponds to brain atrophy and localized edema, from TLE and PNES to the effects of long COVID. While acute NI is healthy and reparative, chronic and sustained microglial cell activation leads to cell death and a perpetual cycle of NI-induced damage. Brain temperature (T_{CRE}) increases occur alongside other neuroinflammatory phenomena, including blood–brain barrier (BBB) disruption, para-arterial cerebrospinal fluid (CSF) influx, infiltration of proinflammatory cells, heightened P-glycoprotein (Pgp) expression, increased peripheral leukocyte extravasation, and edema (elevated extracellular free water as measured by fractional isotropic volume fraction or FISO). While imaging data increasingly support this hypothesis, future studies are needed to establish causal relationships. Figure adapted using Biorender.com, with permission from *Scientific Reports.* (Original citation: Sharma AA, Nenert R, Goodman AM, Szaflarski JP. Brain temperature and free water increases after mild COVID-19 infection. Sci Rep. 2024 Mar 28;14(1):7450. doi: 10.1038/s41598-024-57561-6. PMID: 38548815; PMCID: PMC10978935)

Box 1 Epilepsy as a Disease Model: Focal T_{CRE} Elevations in Seizure-Producing Tissue

Focal brain temperature elevations in patients with localization-related treatment-resistant epilepsy (TRE) may be caused by increased metabolic demands due to (1) chronic, low-level NI within the initial injury site and the seizure onset zone (SOZ) and (2) functionally aberrant and/or structurally damaged neurons in regions implicated in initiating seizures [3, 97–103]. Both phenomena overwhelm metabolic demands and cause an increase in local heat production. This triggers functional, structural, and biochemical alterations in structures within the SOZ and in adjacent and/or connected brain regions within networks involved in seizure propagation and maintenance networks [103, 104]. These cyclical, perpetual phenomena that cause TRE may thus also underlie the pathophysiology of the comorbidities in epilepsy. Created using Biorender.com, with elements from the "Roles of Microglia in Neuroinflammation" template.

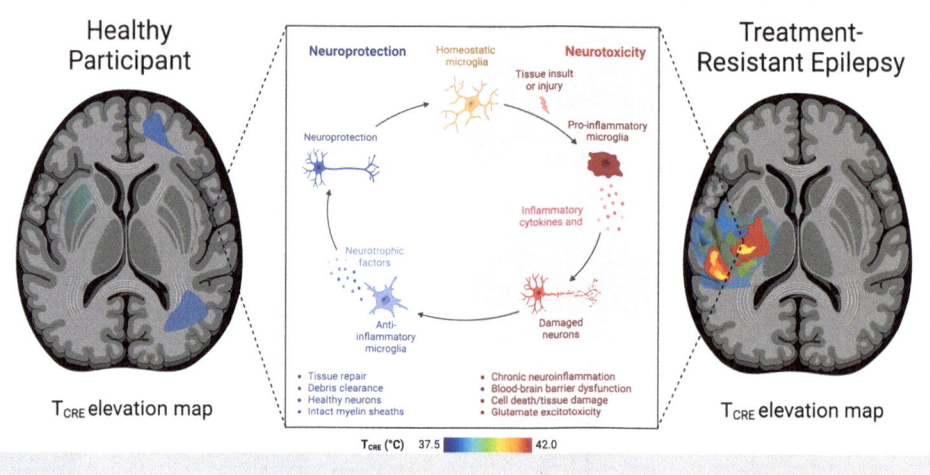

Box 1. Epilepsy as a Disease Model: Focal T_{CRE} Elevations in Seizure-Producing Tissue.

4 Conclusions

With MRSI-t, we can potentially enhance patient outcomes in a range of neurological disorders by increasing our armamentarium of techniques able to noninvasively visualize NI in vivo. This is especially true for disorders such as MS, AD, and mental health disorders where the pathophysiological phenomena are more diffuse and less distinct. To effectively combat neuroinflammatory pathologies, we need to gain a deeper understanding of the complex interplay between NI, disease progression, and response to treatment. For example, a study tracking MRSI-t-based changes in T_{CRE} before and after surgical tissue resection would be particularly valuable in the context of focal epilepsy. Future studies should also examine whether MRSI-t can detect subtle changes in brain temperature over time, particularly in the context of a disease such as MS, where lesions vary both spatially and temporally. In an AD study, understanding the relationship between brain temperature and tau deposition would be paramount, as would longitudinal mapping and monitoring of regional and temporal MRSI-t-based changes in brain temperature as the disease progresses. Given the increased number of metabolites that can be separated using higher power magnets and advanced computational methods, future research could also investigate whether brain temperature mapping in higher power magnets using MRSI-t provides more accurate or detailed results. Ultimately, it is still unclear what precise phenomena establish the causal link between an increase in brain temperature, NI, and the clinical consequences of such chronically activated processes; MRSI-based studies of bioenergetic and immunooxidative abnormalities could be useful in this regard.

References

1. Lucas S-M, Rothwell NJ, Gibson RM (2006) The role of inflammation in CNS injury and disease. Br J Pharmacol 147(Suppl 1):S232–S240

2. Amor S, Puentes F, Baker D, Van Der Valk P (2010) Inflammation in neurodegenerative diseases, pp 154–169

3. Maroso M et al (2010) Toll-like receptor 4 and high-mobility group box-1 are involved in ictogenesis and can be targeted to reduce seizures. Nat Med 16(4):413–419

4. Clausen F et al (2009) Neutralization of interleukin-1beta modifies the inflammatory response and improves histological and cognitive outcome following traumatic brain injury in mice. Eur J Neurosci 30(3):385–396

5. Chen WW, Zhang X, Huang WJ (2016) Role of neuroinflammation in neurodegenerative diseases (review). Mol Med Rep 13(4):3391–3396

6. Tang Y, Le W (2016) Differential roles of M1 and M2 microglia in neurodegenerative diseases. Mol Neurobiol 53(2):1181–1194

7. Davalos D et al (2005) ATP mediates rapid microglial response to local brain injury in vivo. Nat Neurosci 8(6):752–758

8. Yang H, Andersson U, Brines M (2021) Neurons are a primary driver of inflammation via release of HMGB1. Cells 10(10)

9. Ransohoff RM (2016) How neuroinflammation contributes to neurodegeneration. American Association for the Advancement of Science, pp 777–783

10. Weidner LD et al (2018) The expression of inflammatory markers and their potential influence on efflux transporters in drug-resistant mesial temporal lobe epilepsy tissue. Epilepsia 59(8):1507–1517

11. Wang C et al (2023) The effects of microglia-associated neuroinflammation on Alzheimer's disease. Front Immunol 14:1117172

12. Arnett HA et al (2001) TNF alpha promotes proliferation of oligodendrocyte progenitors and remyelination. Nat Neurosci 4(11):1116–1122

13. Streit WJ, Mrak RE, Griffin WST (2004) Microglia and neuroinflammation: a pathological perspective. J Neuroinflammation 1(1):14

14. McNamara NB et al (2023) Microglia regulate central nervous system myelin growth and integrity. Nature 613(7942):120–129

15. Danzer SC (2019) A hit, a hit – a very palpable hit: mild TBI and the development of epilepsy. Epilepsy Curr:153575971985475–153575971985475

16. Pitkänen A et al (2016) In: Laskowitz D, Grant G (eds) Epilepsy after traumatic brain injury, in models of seizures and epilepsy. CRC Press/Taylor and Francis, pp 661–681

17. Tomkins O et al (2008) Blood-brain barrier disruption in post-traumatic epilepsy. J Neurol Neurosurg Psychiatry 79(7):774–777

18. Casillas-Espinosa, P.M., I. Ali, and T.J. O'Brien, Neurodegenerative pathways as

targets for acquired epilepsy therapy development. 2020

19. Ghadery C et al (2017) Microglial activation in Parkinson's disease using [18F]-FEPPA. J Neuroinflammation 14(1):8–8

20. Prokop S, Miller KR, Heppner FL (2013) Microglia actions in Alzheimer's disease, pp 461–477

21. Fan Z et al (2015) Influence of microglial activation on neuronal function in Alzheimer's and Parkinson's disease dementia. Alzheimer's Dement 11(6):608–621.e7

22. Albrecht DS, Granziera C, Hooker JM, Loggia ML (2016) In vivo imaging of human neuroinflammation. ACS Chem Neurosci 7(4):470–483

23. Sharma AA, Szaflarski JP, Sharma A, Szaflarski J (2020) In vivo imaging of neuroinflammatory targets in treatment-resistant epilepsy. Curr Neurol Neurosci Rep 20(4)

24. Jordão MJC et al (2019) Single-cell profiling identifies myeloid cell subsets with distinct fates during neuroinflammation. Science 363(6425)

25. Probert L (2015) TNF and its receptors in the CNS: the essential, the desirable and the deleterious effects. Neuroscience 302:2–22

26. Koepp MJ et al (2017) Neuroinflammation imaging markers for epileptogenesis. Epilepsia 58:11–19

27. French JA (2016) Imaging brain inflammation: if we can see it, maybe we can treat it. Epilepsy Curr 16(1):24–26

28. Nettis MA, Pariante CM (2020) Is there neuroinflammation in depression? Understanding the link between the brain and the peripheral immune system in depression. Int Rev Neurobiol 152:23–40

29. Troubat R et al (2021) Neuroinflammation and depression: a review. Eur J Neurosci 53(1):151–171

30. Kim YK, Na KS, Myint AM, Leonard BE (2016) The role of pro-inflammatory cytokines in neuroinflammation, neurogenesis and the neuroendocrine system in major depression 64:277–284

31. Setiawan E et al (2015) Role of translocator protein density, a marker of neuroinflammation, in the brain during major depressive episodes. JAMA Psychiatry 72(3):268–275

32. Shao F et al (2022) Microglia and neuroinflammation: crucial pathological mechanisms in traumatic brain injury-induced neurodegeneration. Front Aging Neurosci 14:825086

33. Golla SSV et al (2015) Quantification of [18 F]DPA-714 binding in the human brain: initial studies in healthy controls and Alzheimer's disease patients. J Cereb Blood Flow Metab 35(5):766–772

34. Varrone A et al (2015) Positron emission tomography imaging of the 18-kDa translocator protein (TSPO) with [18F]FEMPA in Alzheimer's disease patients and control subjects. Eur J Nucl Med Mol Imaging 42(3):438–446

35. Schuitemaker A et al (2013) Microglial activation in Alzheimer's disease: an (R)-[11C]PK11195 positron emission tomography study. Neurobiol Aging 34(1):128–136

36. Edison P et al (2008) Microglia, amyloid, and cognition in Alzheimer's disease: an [11C](R)PK11195-PET and [11C]PIB-PET study. Neurobiol Dis 32(3):412–419

37. Ghadery C et al (2019) PET evaluation of microglial activation in non-neurodegenerative brain diseases. Curr Neurol Neurosci Rep 19(7):38–38

38. Lockhart A et al (2003) The peripheral benzodiazepine receptor ligand PK11195 binds with high affinity to the acute phase reactant α1-acid glycoprotein: implications for the use of the ligand as a CNS inflammatory marker. Nucl Med Biol 30(2):199–206

39. Schur S et al (2018) Significance of FDG-PET Hypermetabolism in children with intractable focal epilepsy. Pediatr Neurosurg 53(3):153–162

40. Theodore WH (2017) Presurgical focus localization in epilepsy: PET and SPECT. Semin Nucl Med 47(1):44–53

41. Kumlien E et al (2001) PET with 11 C-deuterium-deprenyl and 18 F-FDG in focal epilepsy. Acta Neurol Scand 103(6):360–366

42. Gershen LD et al (2015) Neuroinflammation in temporal lobe epilepsy measured using positron emission tomographic imaging of translocator protein. JAMA Neurol 72(8):882–888

43. Dickstein LP et al (2019) Neuroinflammation in neocortical epilepsy measured by PET imaging of translocator protein. Epilepsia 60(6):epi.15967-epi.15967

44. Owen DR et al (2012) An 18-kDa translocator protein (TSPO) polymorphism explains differences in binding affinity of the PET Radioligand PBR28. J Cereb Blood Flow Metab 32(1):1–5

45. Turkheimer FE et al (2015) The methodology of TSPO imaging with positron emission tomography. Biochem Soc Trans 43(4):586–592

46. Kreisl WC et al (2013) A genetic polymorphism for translocator protein 18 Kda affects

both in vitro and in vivo Radioligand binding in human brain to this putative biomarker of Neuroinflammation. J Cereb Blood Flow Metab 33(1):53–58

47. Sharma AA, Szaflarski JP (2019) Seizing the neuroinflammatory target: the quest continues. Epilepsy Curr 19(6)

48. Vivash L, Obrien TJ (2016) Imaging microglial activation with TSPO PET: lighting up neurologic diseases? J Nucl Med 57(2): 165–168

49. Best L et al (1910) New and old TSPO PET Radioligands for imaging brain microglial activation in neurodegenerative disease

50. Hamelin L et al Early and protective microglial activation in Alzheimer's disease: a prospective study using 18 F-DPA-714 PET imaging

51. Kochanek PM, Jackson TC (2017) The brain and hypothermia—from Aristotle to targeted Temperature management. Crit Care Med 45(2):305–310

52. Wang H et al (2014) Brain temperature and its fundamental properties: a review for clinical neuroscientists. Front Neurosci 8(Sep)

53. Kiyatkin EA (2019) Brain temperature and its role in physiology and pathophysiology: lessons from 20 years of thermorecording. Temperature 6(4):271–333

54. Brain temperature: physiology and pathophysiology after brain injury. Hindawi, pp 13–13

55. Clarke E, Stannard J (1963) Aristotle on the anatomy of the brain. J Hist Med Allied Sci 18(2):130–148

56. Delgado JM, Hanai T (1966) Intracerebral temperatures in free-moving cats. Am J Phys 211(3):755–769

57. Kovalzon VM (1973) Brain temperature variations during natural sleep and arousal in white rats. Physiol Behav 10(4):667–670

58. Hayward JN, Baker MA (1968) Role of cerebral arterial blood in the regulation of brain temperature in the monkey. Am J Phys 215(2):389–403

59. Abrams R, Hammel HT (1964) Hypothalamic temperature in unanesthetized albino rats during feeding and sleeping. Am J Phys 206:641–646

60. Sharma AA et al (2020) Repeatability and reproducibility of in-vivo brain Temperature measurements. Front Hum Neurosci 14: 598435–598435

61. Rzechorzek NM et al (2022) A daily temperature rhythm in the human brain predicts survival after brain injury. Brain 145(6): 2031–2048

62. Zhang Y et al (2020) Reproducibility of whole-brain temperature mapping and metabolite quantification using proton magnetic resonance spectroscopy. NMR Biomed 33(7)

63. Maudsley AA, Goryawala MZ, Sheriff S (2017) Effects of tissue susceptibility on brain temperature mapping. NeuroImage 146:1093–1101

64. Thrippleton MJ et al (2014) Reliability of MRSI brain temperature mapping at 1.5 and 3 T. NMR Biomed 27(2):183–190

65. Rossi S et al (2001) Brain temperature, body core temperature, and intracranial pressure in acute cerebral damage. J Neurol Neurosurg Psychiatry 71(4):448–454

66. Plank JR et al (2022) Brain temperature as an indicator of neuroinflammation induced by typhoid vaccine: assessment using whole-brain magnetic resonance spectroscopy in a randomised crossover study. NeuroImage Clin 35:103053–103053

67. Schiefecker AJ et al Brain temperature but not core temperature increases during spreading depolarizations in patients with spontaneous intracerebral hemorrhage

68. Rumana CS et al (1998) Brain temperature exceeds systemic temperature in head-injured patients. Crit Care Med 26(3):562–567

69. Dietrich W, Bramlett H (2017) Therapeutic hypothermia and targeted temperature management for traumatic brain injury: experimental and clinical experience. Brain Circ 3(4):186–186

70. Dietrich WD et al (1998) Posttraumatic cerebral ischemia after fluid percussion brain injury: an autoradiographic and histopathological study in rats. Neurosurgery 43(3): 585–593

71. Dietrich WD, Atkins CM, Bramlett HM (2017) Is temperature an important variable in recovery after mild traumatic brain injury? Faculty of 1000 Ltd

72. Chatzipanteli K, Alonso OF, Kraydieh S, Dietrich WD (2000) Importance of posttraumatic hypothermia and hyperthermia on the inflammatory response after fluid percussion brain injury: biochemical and immunocytochemical studies. J Cereb Blood Flow Metab 20(3):531–542

73. Dietrich WD, Alonso O, Halley M (1994) Early microvascular and neuronal consequences of traumatic brain injury: a light and electron microscopic study in rats. J Neurotrauma 11(3):289–301

74. Dietrich WD, Alonso O, Halley M, Busto R (1996) Delayed posttraumatic brain hyperthermia worsens outcome after fluid percussion brain injury: a light and electron microscopic study in rats. Neurosurgery 38(3):533–541

75. Sharma HS, Hoopes PJ (2003) Hyperthermia induced pathophysiology of the central nervous system. Int J Hyperth 19:325–354

76. Vosler PS, Logue ES, Repine MJ, Callaway CW (2005) Delayed hypothermia preferentially increases expression of brain-derived neurotrophic factor exon III in rat hippocampus after asphyxial cardiac arrest. Mol Brain Res 135(1–2):1–29

77. Gowda R, Jaffa M, Badjatia N (2018) Thermoregulation in brain injury. Handb Clin Neurol 157:789–797

78. González-Ibarra FP, Varon J, López-Meza EG (2011) Therapeutic hypothermia: critical review of the molecular mechanisms of action. Front Neurol

79. Maudsley AA, Domenig C, Sheriff S (2010) Reproducibility of serial whole-brain MR spectroscopic imaging. NMR Biomed 23(3): 251–256

80. Mueller C et al (2019) Evidence of widespread metabolite abnormalities in Myalgic encephalomyelitis/chronic fatigue syndrome: assessment with whole-brain magnetic resonance spectroscopy. Brain Imaging Behav

81. Maudsley AA et al (2006) Comprehensive processing, display and analysis for in vivo MR spectroscopic imaging. NMR Biomed 19(4):492–503

82. Corbett RJT, Laptook AR, Tollefsbol G, Kim B (1995) Validation of a noninvasive method to measure brain temperature in vivo using 1H NMR spectroscopy. J Neurochem 64(3): 1224–1230

83. Sharma AA, Szaflarski JP (2021) Neuroinflammation as a pathophysiological factor in the development and maintenance of functional seizures: a hypothesis. Epilepsy Behav Rep 16:100496–100496

84. Sharma AA, Nenert R, Goodman A, Szaflarski JP (2023) Concordance between focal brain temperature elevations and focal edema in temporal lobe epilepsy. Epilepsia

85. Zeinali-Rafsanjani B et al (2018) No Title 8(3)

86. Oates M et al (2023) An exploratory study of brain temperature and choline abnormalities in temporal lobe epilepsy patients with depressive symptoms. Epilepsia Open 8(4): 1541–1555

87. Sharma AA, Szaflarski JP (2024) The longitudinal effects of cannabidiol on brain temperature in patients with treatment-resistant epilepsy. Epilepsy Behav 151:109606

88. Verius M, Frank F, Gizewski E, Broessner G (2019) Magnetic resonance spectroscopy thermometry at 3 tesla: importance of calibration measurements. Ther Hypothermia Temp Manag 9(2):146–155

89. Cady EB, D'Souza PC, Penrice J, Lorek A (1995) The estimation of local brain Temperature by in vivo 1H magnetic resonance spectroscopy. Magn Reson Med 33(6):862–867

90. Maudsley AA et al (2021) Advanced magnetic resonance spectroscopic neuroimaging: experts' consensus recommendations. NMR Biomed 34(5):e4309

91. Tzourio-Mazoyer N et al (2002) Automated anatomical labeling of activations in SPM using a macroscopic anatomical parcellation of the MNI MRI single-subject brain. NeuroImage 15(1):273–289

92. Luo T et al (2019) Characterizing structural changes with devolving remyelination following experimental demyelination using high angular resolution diffusion MRI and texture analysis. J Magn Reson Imaging 49(6): 1750–1759

93. Sepehrband F et al (2015) Brain tissue compartment density estimated using diffusion-weighted MRI yields tissue parameters consistent with histology. Hum Brain Mapp 36(9): 3687–3702

94. Mueller C, Sharma AA, Szaflarski JP (2023) Peripheral and central nervous system biomarkers of inflammation in functional seizures: assessment with magnetic resonance spectroscopy. Neuropsychiatr Dis Treat 19: 2729–2743

95. Sharma A, Nenert R, Goodman A, Szaflarski JP (2024) Brain temperature and free water increases after mild COVID-19 infection. Sci Rep 14(1):7450

96. Annink KV et al (2020) Brain temperature of infants with neonatal encephalopathy following perinatal asphyxia calculated using magnetic resonance spectroscopy. Pediatr Res:1–6

97. Jehi L (2018) The epileptogenic zone: concept and definition. American Epilepsy Society, pp 12–16

98. Ren Y et al (2019) Transient seizure onset network for localization of epileptogenic zone: effective connectivity and graph theory-based analyses of ECoG data in temporal lobe epilepsy. J Neurol 266(4):844–859

99. Abela E et al (2014) Neuroimaging of epilepsy: lesions, networks, oscillations. Urban und Vogel GmbH, pp 5–15

100. Bernhardt BC, Bonilha L, Gross DW (2015) Network analysis for a network disorder: the emerging role of graph theory in the study of epilepsy. Academic Press, pp 162–170

101. Bernhardt BC, Bernasconi N, Kim H, Bernasconi A (2012) Mapping thalamocortical network pathology in temporal lobe epilepsy. Neurology 78(2):129–136

102. Englot DJ et al (2010) Impaired consciousness in temporal lobe seizures: role of cortical slow activity. Brain 133(Pt 12):3764–3777

103. Holmes GL (2015) Cognitive impairment in epilepsy: the role of network abnormalities. Epileptic Disord 17(2):101–116

104. Klein P et al (2018) Commonalities in epileptogenic processes from different acute brain insults: do they translate? Epilepsia 59(1):37–66

Chapter 10

Magnetic Resonance Spectroscopy and Its Application of Mild Traumatic Brain Injury

Sihong Huang and Jun Liu

Abstract

The high incidence of mild traumatic brain injury (mTBI) and associated post-concussion symptoms, such as headache and cognitive deficits, have garnered significant attention from researchers worldwide. They are consistently searching for concrete evidence that can promptly indicate the involvement of nervous system. Magnetic resonance spectroscopy (MRS), a noninvasive technique derived from magnetic resonance imaging (MRI), provides the capability to investigate brain metabolites as biomarkers of pathophysiological changes in vivo following mTBI. In this chapter, we begin by providing a brief overview of the background of mTBI. Subsequently, we delve into the detailed processes involved in MRS data, ranging from data collection to result presentation.

Key words Mild traumatic brain injury, MRI, Magnetic resonance spectroscopy, Metabolites, Pathophysiological biomarkers

1 Introduction

More than 54 million people suffer a traumatic brain injury (TBI) worldwide each year [1]. The mild type of TBI (mTBI) is the most common one, accounting for 80% of TBI patients [2, 3]. However, it is well known that traditional MRI or CT scans often yield negative results in mTBI cases. Despite this, the widespread prevalence of persistent post-concussion symptoms, including headache and cognitive impairments, highlights the long-term impact of mTBI on the central nervous system. These observations have spurred scientists to consistently seek definitive evidence that can promptly indicate nervous system involvement.

Diffuse axonal injury (DAI), the abnormal release of excitatory amino acids and neurotransmitters, disruptions in glycolytic metabolism, as well as neurometabolic cascade following mTBI, can be detected using a noninvasive imaging technique known as magnetic resonance spectroscopy (MRS), either directly or indirectly [4]. MRS, a noninvasive sequence derived from magnetic

Daichi Sone (ed.), *Molecular Imaging for Brain Diseases*, Neuromethods, vol. 222, https://doi.org/10.1007/978-1-0716-4494-2_10,

resonance imaging (MRI), allows the examination of brain chemistry as biomarkers of pathophysiology in vivo. It involves spectral peaks that correspond to the signal intensity emitted by protons within specific metabolites. Each peak signifies the concentration of the corresponding metabolite, as it is directly proportional to that concentration. The position on the axis for a particular metabolite is indicated by its chemical shift concerning a known reference standard, typically expressed in parts per million (ppm). Distinguishing metabolites with very similar chemical shifts can be challenging, so it is common to report composite metabolites. Proton MRS is typically used to detect well-known metabolites, such as N-acetylaspartate (NAA) (2.01 ppm), creatine (Cr) (3.03 ppm), choline (Cho) (3.20 ppm), the combination (Glx) of glutamate (Glu) and its precursor glutamine (Gln) (2.05–2.5 ppm and 3.7 ppm), myo-Inositol (mI) (3.56 ppm), and lactate (Lac) (1.31 ppm). These metabolites serve as markers of neuronal health, brain energy metabolism, membrane synthesis or repair, inflammation or demyelination, excitotoxicity reactions, glial proliferation, and anaerobic glycolysis, respectively. Moreover, two-dimensional MRS techniques like localized correlated spectroscopy have been employed to detect macromolecules in proton MRS [5]. Additionally, phosphorus spectroscopy has been utilized to assess adenosine triphosphate (ATP) metabolism and intracellular pH changes following mTBI. This approach can reflect disturbances in energy metabolism and the accumulation of lactic acid due to intracellular and intracellular ion imbalance after mTBI [6].

In previous MRS studies involving mTBI patients, it has been consistently observed that the concentration of NAA is significantly reduced when compared to control groups. However, the results for other metabolites have shown varying directions. These discrepancies can be attributed to several factors, including the inherent heterogeneity among mTBI patients, differences in occupation, causes, stages, severities and the number of injuries, variations in control groups, variations in the assessment of clinical features, and the selection of regions of interest (ROIs) in different studies. One strategy to address these discrepancies is to significantly increase the sample size, enabling fine subgroup divisions and stringent control of single variables. Additionally, differences in scanner types and data acquisition methods are critical factors contributing to different results. The Enhancing NeuroImaging Genetics through Meta-Analysis (ENIGMA) MRS working group has recommended the use of single voxel MRS as the standardized sequence across multi-center scanners to obtain robust findings related to mTBI [7]. Regrading data processing, some studies have utilized software provided by vendors. However, the LCModel is the most widely adopted software for extracting concentration data from various metabolites, as it has been specifically developed for this purpose.

2   Methods

2.1   Participants

The inclusion criteria for mTBI patients were determined based on the guidelines provided by the WHO's Collaborating Center for Neurotrauma Task Force [8]. Patients meeting the following criteria were included: (1) a Glasgow Coma Scale score of 13–15; (2) presence of any of the following: (a) confusion or disorientation, (b) loss of consciousness (LOC) ≤ 30 min, (c) posttraumatic amnesia <24 h, (d) transient neurological abnormalities (focal signs or seizure), or (e) intracranial lesion not requiring surgery; and (3) evaluation within 14 days after the onset of mTBI as the acute group as the baseline. We also recruited these patients after 3 months, 6 months, and 1-year follow-up. Exclusion criteria included (1) presence of severe psychiatric disease, severe somatic disease, or drug abuse; (2) history of complicated mild, moderate, or severe TBI or other diseases associated with brain pathology; (3) structural abnormality; and (4) contraindications for magnetic resonance imaging (MRI). In most previous studies involving community mTBI patients, the control group consisted of age-, sex-, and education-matched healthy controls (HCs). These HCs were recruited simultaneously with mTBI patients and had to meet the same exclusion criteria as the mTBI patients. Another choice for controls is orthopedic injury (OI) patients, which allows for the study of metabolites specific to concussion [9]. However, there is debate regarding the advantages of using OI patients as controls, as some argue that OI may not provide clear advantages [10, 11]. Some studies suggest that a comparison among mTBI patients, HCs, and OI groups may be a better design to assess the impact of mTBI. Furthermore, non-concussed athletes are often used as the control group when comparing them to athletes with concussions. Similarly, military personnel without a history of brain injury are frequently chosen as the control group for military mTBI patients.

2.2   MRI Acquisition

In earlier MRS studies involving mTBI patients, 1.5 Tesla (T) scanners were commonly used. However, in recent years, 3 T MRI scanners have become the mainstream, and even 7 T scanners have been utilized in some previous studies [12, 13]. The shift from 1.5 T to 3 T scanners represents an advancement in MRI technology. Earlier MRS studies typically employed 8- or 12-channel head coils for data acquisition. In more recent studies, the 32-channel head coil has become the primary choice for acquiring data. The use of a 32-channel head coil allows for improved signal quality and data acquisition. In our research, a 3 T scanner with a 32-channel head coil was utilized. The imaging protocols included clinical diagnostic sequences, such as T1-weighted imaging, T2-weighted imaging, and fluid-attenuated inversion recovery (FLAIR).

Additionally, a high-resolution structural MRI sequence known as magnetization prepared rapid acquisition gradient echo (MPRAGE) and susceptibility-weighted imaging (SWI) were employed to thoroughly assess and rule out any structural abnormalities. The interpretation of these imaging sequences was carried out by at least two experienced neuroradiologists with several years of expertise in neuroimaging. In cases where there was a discrepancy in their assessments, a consensus was reached through discussion and collaboration. This rigorous approach helps ensure the accuracy and reliability of the imaging data.

In the context of mTBI, two primary pulse sequences have been employed for MRS studies: single voxel spectroscopy (SVS) and multivoxel MR spectroscopic imaging (MRSI). SVS is known for its simplicity and speed, making it an efficient choice for data acquisition. However, it has a limitation—it can only examine one or a few specific locations sequentially in each acquisition. This sequential approach can be time-consuming when assessing multiple brain regions. On the other hand, MRSI allows for the simultaneous acquisition of data from multiple brain regions, offering a more comprehensive view of metabolite concentrations. However, it comes with challenges, including significant variability between subjects and across different MRI scanners [14]. Additionally, MRSI often yields a smaller signal-to-noise ratio (SNR) due to the smaller voxel sizes used [15]. Two common localization schemes in MRS are the Point RESolved Single voxel (PRESS) and Stimulated Echo Acquisition Mode (STEAM) sequences. These localization schemes play a crucial role in determining the accuracy of metabolite quantification. A meta-analysis conducted among mixed TBI patients found differences between subgroups based on localization schemes [16]. The STEAM subgroup exhibited decreased NAA, increased Cho/Cr, and increased Cho, whereas these trends were not observed in the PRESS subgroup. This suggests that STEAM may have certain advantages in evaluating traumatic brain injuries. However, it is important to note that direct comparison results of these two localization schemes are often challenging to obtain because they are typically studied in different groups of patients. The ENIGMA recommended using PRESS as the localization scheme and SVS as pulse sequence for TBI studies. The choice was made based on the higher SNR, availability on all commercial MRI scanners, and frequent use in routine clinical settings. This standardization aims to seek robust effects of mild TBI across multi-center and multi-site datasets. Furthermore, ENIGMA recommended using a 3.0 T scanner with a repetition time (TR) of 2000 ms, echo time (TE) of a minimum of 30–35 ms, and a number of averages (NAV) of 128. An acquisition without water suppression with 16 NAV was conducted. These parameters were selected to ensure data consistency and reliability. It is also important to note that an identical non-water-suppressed acquisition with at least eight averages is needed, especially for the absolute concentration calculation of metabolites. If a non-water-suppressed sequence was not collected, phantom replacement or relative concentration methods can be used as alternatives. These standardized methods were employed in our study to ensure data quality and comparability with specified parameters: repetition time of 2000 ms, echo time of 30 ms, NAV of 128, and number of points of 1024. An acquisition without water suppression with 16 NAV conducted. All participants were positioned in a supine position with foam padding to minimize head motion during data acquisition.

The selection of ROIs is a critical aspect that can significantly influence the results of studies on mTBI. The choice of ROIs is typically based on prior knowledge and an understanding of the specific pathological changes associated with mTBI. Given that DAI is the primary and most significant pathological change following brain injury [4], studies often focus on white matter (WM) ROIs. Among mTBI patients, the corpus callosum (CC), frontal WM, and centrum semiovale are the three most commonly selected ROIs in previous research. These regions are chosen because they are frequently affected by mTBI-related damage. Gray matter (GM) also experiences noticeable pathological changes following mTBI. Alterations of cerebral blood flow, as well as correlation between cortical areas and cognitive or emotional performance, make GM an important ROI choice. Regions such as the prefrontal cortex (PFC), especially dorsolateral PFC (DLPFC), cingulate cortex, and the initial motor cortex (M1), are frequently examined due to their relevance to mTBI. In MRS studies, particularly in SVS studies, relatively large voxels are often used to capture sufficient signal. To obtain accurate metabolite results, it is essential to exclude unwanted issue components that can impact the MRS data, such as cerebrospinal fluid, skull, and lipids. Moreover, the combination of different tissues types within the ROIs, including WM and GM, can be a confounding factor in the results. Different tissue types may exhibit distinct metabolite concentrations. For example, concentrations of NAA, Cr, and mI tend to be higher in GM compared to WM, while Cho concentrations are notably lower in GM [17]. Additionally, the concentrations of NAAG in WM are significantly greater than those in GM [18]. However, it is often unrealistic to expect an ROI in an SVS MRS study to contain only a single tissue component, especially when GM is selected as the ROI. Therefore, it is essential to perform comprehensive tissue corrections that account for the presence of different tissue components within the ROIs. Furthermore, it is worth noting that different subregions within the CC may response differently to brain injury stress, and their metabolite levels may vary [19]. Careful and strict selection of ROIs is crucial to obtaining meaningful results. In line with ENGIMA's recommendations, the frontal WM of 8*8*8 mm^3 was chosen as the ROI for detecting metabolite changes after mTBI in our study. This standardized approach ensures data consistency and comparability across studies.

2.3 Data Processing   As outlined in the expert consensus recommendations of Jamie Near and his colleagues [20], the processing of MRI data can be divided into three distinct steps: preprocessing, analysis, and quantification. These steps are crucial for accurately and rigorously evaluating MRI data in various research and clinical applications.

Data preprocessing is the critical first step in the analysis of MRS data. According to the consensus recommendations by Jamie Near and his colleagues, it is strongly recommended to use data formats that preserve individual transients and receiver channels. Various experimental imperfections, such as eddy currents, subject motion, or scanner drift, can potentially degrade MRS data quality. Eddy current effects can be corrected by addressing unwanted short-lived fluctuation in the B0-field caused by rapid gradient switching. One common approach to dealing with this issue is to acquire an unsuppressed water spectrum. To detect gross motion artifacts, a practical method is to acquire quick localizer images immediately before and after the MRS scan and compare the position of the anatomy of interest between these two scans. However, subjects who cooperate well and maintain head fixation typically experience minimal motion artifacts that have little impact on the data. Practical method is to acquire quick localizer images immediately before and after the MRS scan and compare the position of the anatomy of interest between these two scans. Temporal drift in the main magnetic field of the MRI scanner can occur due to heating and cooling of ferromagnetic passive shim elements in thermal contact with the gradient coils. Physiological or bulk motion during the MRS scan can also introduce additional frequency and phase offsets. If not corrected, these frequency and phase drifts can lead to broadening of spectral peaks, reduction in signal-to-noise ratio (SNR), and line shape distortion. While scanner software can perform online drift correction, it is recommended to apply additional offline correction, as online correction may address frequency but not phase drifts. Retrospective correction of frequency and phase drifts can be achieved by tracking the frequency and phase of the residual water peak using separately acquired navigator echoes or individual metabolite peaks. Additionally, multi-dimensional raw data from multiple coil channels should be integrated into a one-dimensional spectrum that is amenable to analyzed. Combining signals from RF coils or performing signal averaging are common preprocessing methods in this regard. These steps are crucial for ensuring the quality and integrity of MRS data before subsequent analysis and quantification.

The second step in the analysis of MRS data involves estimating MRS peak areas, and this can be accomplished using three common approaches: (1) Linear combination model fitting: This method involves fitting the MRS data using a linear combination model. It models the full spectral contribution of each metabolite by using basis spectra. Linear combination model fitting is the most popular and widely recommended method for most in vivo MRS applications due to its robustness and accuracy (2). Peak fitting: In this approach, individual peaks of interest within a spectrum are selected, and each peak is fitted using a simple line shape model function. This method focuses on fitting specific peaks of interest rather than modeling the entire spectral contribution of each metabolite. (3) Peak integration: Peak integration estimates a metabolite's signal intensity by calculating the area under its peak in the frequency domain. This method quantifies metabolite concentrations based on the integral of the spectral peaks. It offers a comprehensive way to quantify metabolite concentrations by modeling the full spectral contributions of each metabolite. However, the choice of method may depend on the specific research question and the nature of the MRS data being analyzed.

The third step in the analysis of ^{1}H-MRS data is quantification. Quantification involves converting the ^{1}H-MRS signals obtained from the brain into metabolite concentrations. This process is achieved by comparing the metabolite signals to either internal or external chemical concentration reference or to an externally synthesized signal. The internal references utilize signals from within the brain itself. Common internal references include the tissue water signal or a combination of signals from metabolites like Cr, Cho, and NAA within the metabolite spectrum. By comparing the metabolite signals to these internal references, researchers can estimate metabolite concentrations relative to the chosen reference. External standards involve the use of a chemical of known concentration in a solution. This standard is either positioned close to the subject's head during the MRS scan or scanned before or after the subject is scanned (known as "phantom replacement"). By comparing the metabolite signals to this external standard, it is possible to determine the absolute concentration of metabolites.

All clinical scanners offer spectral processing packages for visual interpretation and metabolite quantification. However, several toolboxes were applied in previous researches. And the LCModel [21], which can automatically quantitate metabolites in vivo proton MRS, is the most popular software package of previous MRS studies among mTBI patients.

In our study, the preprocessing of the data was conducted using Siemens software before exporting the overall average spectrum for further analysis. The absolute concentration of NAA, Cho, Cr, and Glx and concentration ratio relative to Cr were estimated using

LCModel package (http://lcmodel.com/lcmodel.shtml). To assess the quality of spectral curve fitting by LCModel and the reliability of estimated metabolite concentrations, several parameters were considered: (1) Residuals: Residuals were used to evaluate the SNR in the spectra (*see* **Note 1**). An adequate SNR is crucial for accurate spectral analysis (2). Full width at half-maximum (FWHM): FWHM is a rough estimate of the linewidth in the in vivo spectrum. (3) Cramér-Rao lower bound (CRLB): CRLB was employed to assess the accuracy of metabolite concentrations. It is expressed as a percentage standard deviation (%SD) of the estimated concentration. The %SD considers various factors, including resolution and noise level, to provide a more comprehensive measure of the reliability of the estimates. The FWHM <0.10 and SNR <50 were used for the standard of quality control [22, 23]. And voxels with %SD $<20\%$ were included in our analysis, ensuring that the data used for further investigation met the quality standards. It is important to note that no single metric can fully characterize data quality, and these parameters collectively provide a useful guide to assess the reliability of your MRS data.

In order to account for variations in tissue composition within the selected voxels and to adjust metabolite ratio measurements, voxel-wise tissue probability values were calculated using each participant's T1-weighted anatomical scan. This calculation was performed with FMRIB's Automated Segmentation Tool (FAST) [24]. FAST is a tool designed to segment anatomical scans into different tissue types, such as gray matter, white matter, and cerebrospinal fluid (CSF). For each voxel, FAST estimates a partial volume (PVE), which represents the proportion of a specific tissue type within that voxel. These PVE values were included in our metabolite-based statistical analyses as covariates.

2.4 Statistical Analysis

The demographic characteristics and neuropsychological data were subjected to statistical analysis using IBM SPSS Statistics 24.0 for Windows. Two-sample t-tests, chi-square tests, Kruskal–Wallis tests, and Wilcoxon tests were performed for age, gender, education, and neuropsychological tests. Three neuropsychological tests for cognitive function were conducted in our research: the Digit Symbol Substitution Test (DSST) and the Trail Making Test (TMT) A & B. Differences in metabolite levels were analyzed using independent two samples t-tests for cross-sectional comparison and paired two sample t-tests for longitudinal comparison. In addition to the statistical methods for normal distribution data, nonparametric tests were conducted for cross-sectional or longitudinal test among non-normal distribution data. Correlations between metabolite levels and clinical characteristics (e.g., injury and scan interval, neuropsychological tests as we mentioned above) were evaluated using bivariate correlation or partial correlation. Statistical significance was definite as $p < 0.05$. LCModel could generate the one-page output including the concentration table and plot, from which we could apply in our manuscript for result representing (*see* **Note 2**).

3 Notes

1. Residuals represent the discrepancies between the actual data values and the values predicted by the fit. In the one-page output, these residuals are typically displayed at the top of the plot. In a good analysis, these residuals should exhibit a random and even distribution around zero. If the residuals show a clear nonrandom pattern, it may indicate a problem with the analysis and could be grounds for rejection. The randomness of the residuals is an important indicator of the quality of the fit.

2. The thin curve in the main plot of the one-page output represents the baseline. The baseline is a critical component of the analysis because it must be flexible enough to account for various unpredictable complications. These complications can include artifacts in the data, substances not present in the basis set of model metabolite spectra, substances that might be in the basis set but have abnormally short T2 times, inaccuracies in the simulated models for highly variable lipid and macromolecule signals, and incomplete water suppression. In an acceptable analysis, the baseline should exhibit a reasonable level of flexibility. However, if the baseline appears erratic, particularly with large dips that deviate significantly from the data, this suggests problems with the analysis and should be the cause for potential rejection. The quality and appropriateness of the baseline are crucial for accurately estimating metabolite concentrations, and deviations from expected baseline behavior can signal issues with the analysis.

References

1. Bazarian JJ, Biberthaler P, Welch RD et al (2018) Serum GFAP and UCH-L1 for prediction of absence of intracranial injuries on head CT (ALERT-TBI): a multicentre observational study. Lancet Neurol 17:782–789

2. Huang SH, Li MJ, Yeh FC, Huang CX, Zhang HT, Liu J (2023) Differential and correlational tractography as tract-based biomarkers in mild traumatic brain injury: a longitudinal MRI study. NMR Biomed 36:e4991

3. Jiang JY, Gao GY, Feng JF et al (2019) Traumatic brain injury in China. Lancet Neurol 18:286–295

4. Capizzi A, Woo J, Verduzco-Gutierrez M (2020) Traumatic brain injury. Med Clin North Am 104:213–238

5. Velan SS, Lemieux SK, Raylman RR et al (2007) Detection of cerebral metabolites by single-voxel-based PRESS and COSY techniques at 3T. J Magn Reson Imaging 26:405–409

6. Wilke S, List J, Mekle R et al (2017) No effect of anodal transcranial direct current stimulation on gamma-aminobutyric acid levels in patients with recurrent mild traumatic brain injury. J Neurotraum 34:281–290

7. Bartnik-Olson BL, Alger JR, Babikian T et al (2021) The clinical utility of proton magnetic resonance spectroscopy in traumatic brain injury: recommendations from the ENIGMA MRS working group. Brain Imaging Behav 15:504–525

8. Holm L, David Cassidy J, Carroll L, Borg J (2005) Summary of the WHO collaborating Centre for neurotrauma task force on mild traumatic brain injury. J Rehabil Med 37:137–141

9. Ware AL, Yeates KO, Geeraert B et al (2022) Structural connectome differences in pediatric mild traumatic brain and orthopedic injury. Hum Brain Mapp 43:1032–1046

10. Beauchamp MH, Landry-Roy C, Gravel J, Beaudoin C, Bernier A (2017) Should young children with traumatic brain injury be compared with community or orthopedic control participants? J Neurotraum 34:2545–2552

11. Mathias JL, Dennington V, Bowden SC, Bigler ED (2013) Community versus orthopaedic controls in traumatic brain injury research: how comparable are they? Brain Inj 27:887–895

12. Kontos AP, Van Cott AC, Roberts J et al (2017) Clinical and magnetic resonance spectroscopic imaging findings in veterans with blast mild traumatic brain injury and post-traumatic stress disorder. Mil Med 182:99–104

13. Hetherington HP, Hamid H, Kulas J et al (2014) MRSI of the medial temporal lobe at 7 T in explosive blast mild traumatic brain injury. Magn Reson Med 71:1358–1367

14. Sabati M, Sheriff S, Gu M et al (2015) Multi-vendor implementation and comparison of volumetric whole-brain echo-planar MR spectroscopic imaging. Magn Reson Med 74:1209–1220

15. Davitz MS, Gonen O, Tal A, Babb JS, Lui YW, Kirov II (2019) Quantitative multivoxel proton MR spectroscopy for the identification of white matter abnormalities in mild traumatic brain injury: comparison between regional and global analysis. J Magn Reson Imaging 50:1424–1432

16. Brown M, Baradaran H, Christos PJ, Wright D, Gupta A, Tsiouris AJ (2018) Magnetic resonance spectroscopy abnormalities in traumatic brain injury: a meta-analysis. J Neuroradiol 45:123–129

17. Krukowski P, Podgorski P, Guzinski M, Szewczyk P, Sasiadek M (2010) Analysis of the brain proton magnetic resonance spectroscopy – differences between normal grey and white matter. Pol J Radiol 75:22–26

18. Choi C, Ghose S, Uh J et al (2010) Measurement of N-acetylaspartylglutamate in the human frontal brain by 1H-MRS at 7 T. Magn Reson Med 64:1247–1251

19. Johnson B, Zhang K, Gay M et al (2012) Metabolic alterations in corpus callosum may compromise brain functional connectivity in MTBI patients: an 1H-MRS study. Neurosci Lett 509:5–8

20. Near J, Harris AD, Juchem C et al (2021) Preprocessing, analysis and quantification in single-voxel magnetic resonance spectroscopy: experts' consensus recommendations. NMR Biomed 34:e4257

21. Provencher SW (1993) Estimation of metabolite concentrations from localized in vivo proton NMR spectra. Magn Reson Med 30:672–679

22. Alosco ML, Tripodis Y, Rowland B et al (2020) A magnetic resonance spectroscopy investigation in symptomatic former NFL players. Brain Imaging Behav 14:1419–1429

23. Veeramuthu V, Seow P, Narayanan V et al (2018) Neurometabolites alteration in the acute phase of mild traumatic brain injury (mTBI). Acad Radiol 25:1167–1177

24. Zhang Y, Brady M, Smith S (2001) Segmentation of brain MR images through a hidden Markov random field model and the expectation-maximization algorithm. Ieee T Med Imaging 20:45–57

Chapter 11

Glutamate and Proton MR Spectroscopy in Schizophrenia

Teruki Koizumi and Ariel Graff-Guerrero

Abstract

The glutamate hypothesis, a theory that emerged in 1980, suggests that its receptors are promising drug targets for treating treatment-resistant schizophrenia. Research in this field has been conducted through various methods, including genetics, postmortem brain studies, animal models, and neuroimaging. Notably, proton magnetic resonance spectroscopy (^{1}H-MRS) has become one of the most reliable techniques not only for identifying future drug targets but also for elucidating the pathophysiology of this disease. To date, several meta-analyses on ^{1}H-MRS studies in schizophrenia have reported lower glutamate levels in the medial prefrontal cortex (mPFC) and the anterior cingulate cortex (ACC) in the patient group compared to healthy controls. These findings may explain potential neurocircuits observed in animal models of schizophrenia, yet more data are needed for a comprehensive understanding. The development of drugs targeting mGluR2/3, a metabotropic glutamate receptor, showed promise by passing phase 2 clinical trials but ultimately failed in phase 3. Despite these setbacks, the exploration of the glutamatergic system remains a central focus for the future of schizophrenia research, underscoring the continued need for data accumulation.

Key words Schizophrenia, Glutamate, GABA, ^{1}H-MRS, NMDA

1 Introduction

1.1 Schizophrenia

Schizophrenia is a serious mental disorder characterized by symptoms including positive, negative, and cognitive symptoms, often leading to major functional deterioration. Historically, this disease was classified into subtypes including catatonic type, disorganized type, paranoid type, and residual type according to the predominant symptoms. However, in 2013, the *Diagnostic and Statistical Manual of Mental Disorders*, Fifth Edition (*DSM-5*) working group recommended that these subtypes should not be included in *DSM-5*. This significant modification marked a departure from the traditional diagnostic system [1].

However, despite numerous efforts, clear biomarkers for diagnosing this disease have not yet been established, making diagnosis reliant on subjective reports from patients and assessments by clinicians.

Daichi Sone (ed.), *Molecular Imaging for Brain Diseases*, Neuromethods, vol. 222, https://doi.org/10.1007/978-1-0716-4494-2_11,

Schizophrenia affects 0.5–1.0% of the world's population and imposes a substantial burden on society worldwide [2–4]. This disease is associated with significantly higher morbidity and mortality rates compared to the general population, and with the mean age at death being 15–21 years younger [5], alongside notable personal and societal expenses. In fact, according to the Global Burden of Disease Study 2017, schizophrenia ranks as the 15th leading cause among the top 20 years lived with disability (YLDs) for both males and females.

Furthermore, the treatment of schizophrenia faces considerable challenges, particularly in addressing cognitive and negative symptoms. Antipsychotics, which block dopamine D_2 receptors, are efficacious for treating positive symptoms and thought disorders. However, antipsychotics arguably provide little to no benefit for negative symptoms and cognitive impairment [6].

Schizophrenia is also a heterogeneous disorder, with a substantial proportion of patients showing partial or no response to first-line antipsychotic treatments [7]. It is estimated that approximately 25–30% of patients treated with clozapine fail to show an adequate response, a condition known as treatment-resistant schizophrenia (TRS) [8–10]. Among these TRS cases, 40–70% exhibit ultra-treatment resistance (URS), showing no improvement even after undergoing second-line treatment with the antipsychotic medication clozapine [11]. Consequently, this means that 12–20% of all individuals diagnosed with schizophrenia (SCZ) fall into the URS category [10].

It is clear that the dopamine hypothesis alone does not fully account for the complexity of schizophrenia, particularly concerning the issue of treatment resistance. Specifically, dopamine D2 blockers have limited efficacy in addressing negative symptoms and cognitive deficits.

1.2 Glutamatergic Hypothesis and NMDA Receptor Hypofunction Theory

The glutamate hypothesis of schizophrenia emerged in 1980 [12]. Glutamate, the primary excitatory neurotransmitter in the brain, plays a crucial role in the disorder. Glutamate is utilized by glutamatergic neurons and is involved in 60–80% of synaptic transmissions, highlighting its significance in brain function. This glutamate system is promising for explaining the diverse neural mechanisms underlying the heterogeneity of schizophrenia [13].

Glutamate interacts with two main receptor types: ionotropic glutamate receptors (iGluRs) and metabotropic glutamate receptors (mGluRs) [14]. Glutamate binding activates an ion channel pore in iGluRs, whereas in mGluRs, the activation of ion channels on the plasma membrane occurs indirectly via a signaling cascade. iGluRs are divided into three receptor subtypes: N-methyl-D-aspartate (NMDA) receptors, α-amino-3-hydroxy-5-methyl-4-isoxazolepropionicacid (AMPA) receptors, and kainate receptors. mGluRs are categorized into three groups: group 1 (mGluR1 and mGluR5), group 2 (mGluR2 and mGluR3), and group 3 (mGluR4, mGluR6, mGluR7, and mGluR8), encompassing a total of eight identified subtypes that initiate G-protein-coupled signal transduction. NMDA receptors, which are extensively expressed across the brain cortex, play a pivotal role in functions that are known to be compromised in schizophrenia patients, including synaptic plasticity, learning, and memory formation [15].

The NMDA hypofunction theory of schizophrenia is based on two findings: (1) Administration of NMDAR antagonists, such as ketamine, phencyclidine (PCP), dizocilpine (MK-801) induce a psychotic state in healthy volunteers and exacerbate psychotic symptoms in patients with schizophrenia [16, 17] and (2) The presence of anti-NMDA receptor encephalitis, an autoimmune condition that manifests symptoms like schizophrenia [18, 19]. Additionally, NMDAR antagonists closely mimic the symptoms of schizophrenia more accurately than any model based on dopamine dysregulation. This includes not only the positive/psychotic symptoms of the disease induced by dopamine-increasing drugs like amphetamine but also other core symptoms of schizophrenia, such as cognitive impairments and negative symptoms [16, 20].

Considering that NMDA receptors are located on brain circuits that regulate dopamine release, this suggests that the dopaminergic deficits in schizophrenia may also be secondary to underlying glutamatergic dysfunction [21].

Notably, in the maternal immune activation (MAM) rodent model of schizophrenia, the transplantation of γ-aminobutyric acid (GABA)ergic precursor cells into the ventral subiculum of the hippocampus was found to normalize hyperactivity and rejuvenate the downstream dopamine system. These findings provide significant insights into potential therapeutic strategies [22].

In addition to the evidence related to NMDA receptor antagonism, postmortem, genetic research, and neuroimaging studies have indicated that several components of the glutamatergic signaling system are altered in patients with schizophrenia. These aspects will be discussed in the following topic.

2 Literature Review

2.1 Neuroimaging and MRS Findings

Glutamate function in patients with schizophrenia has been investigated in vivo using neuroimaging techniques, including proton magnetic resonance spectroscopy (^{1}H-MRS), positron emission tomography (PET), and single-photon emission computed tomography (SPECT) [15, 23]. ^{1}H-MRS has been used to measure the levels of glutamate (Glu), glutamine (Gln), and their combination, referred to as Glx, to elucidate glutamatergic functioning. Glx describes the combined spectra of Glu and Gln because of their overlapping peaks on the ppm axis, which appear between 2.1 and 2.5 ppm and between 3.72 and 3.82 ppm [24–26]. It has been reported that field strengths of 4 Tesla (T) or higher can improve the accuracy of separating glutamate from glutamine signals [27].

^{1}H-MRS enables noninvasive measurement of small molecules, including the chemical composition of tissues, energy metabolism, and neurotransmitter levels in the living human brain, by detecting differences in resonance frequencies [26]. It primarily utilizes the atomic nuclei of hydrogen (^{1}H), phosphorus (^{31}P), and carbon (^{13}C) in MRS studies due to their magnetic properties. Among these, ^{1}H-MRS is the most frequently applied measurement as the ^{1}H nucleus forms the most sensitive and most abundant, naturally occurring nucleus [28, 29]. The higher sensitivity of ^{1}H-MRS makes it better suited for examining specific small brain regions, as opposed to ^{31}P-MRS.

This technique is notable for directly measuring brain molecules like glutamate without using ionizing radiation. It aligns closely with magnetic resonance imaging (MRI) by detecting signals from hydrogen atoms, with the potential to also assess other nuclei, including carbon and nitrogen. MRS is particularly adept at analyzing predetermined regions of interest and is progressively being developed to allow for a comprehensive biochemical profile of the entire brain. In MRS, each proton's position on the chemical shift axis uniquely identifies its specific chemical environment, facilitating a precise mapping and analysis of the brain's chemical composition. This precision offers vital insights into various neurological conditions [15].

The presence of a hyperglutamatergic state across various brain areas in patients with schizophrenia has been empirically confirmed and replicated through numerous MRS studies. These findings highlight the crucial role of glutamatergic dysfunction in the disorder. Preclinical data further underscore the significance of this target by demonstrating that experimentally induced NMDAR hypofunction leads to an increased firing rate of glutamatergic neurons in animal models [30]. This condition mirrors psychosis-like behavioral phenotypes [31–33] and results in glutamatergic excess in healthy human subjects [34–36].

PET and SPECT studies in healthy subjects showed an impact of NMDA blockade (using ketamine) on dopaminergic indices. This provides initial evidence supporting the glutamate/NMDA hypothesis of schizophrenia. Subsequent advancements in PET and SPECT imaging technology have enabled direct in vivo measurements of glutamatergic indices, crucial for transforming preclinical insights and clinical observations into effective treatments [23].

Despite the rapid progress in the preclinical identification of PET/SPECT ligands targeting different glutamatergic elements, to date, only a few clinical studies have been conducted to assess direct measurements of glutamatergic elements in vivo. SPECT studies in healthy volunteers showed that ketamine infusion led to decreases in NMDAR availability. Similarly, research involving individuals with schizophrenia has indicated reduced NMDAR availability in the hippocampus of drug-free patients. Collectively, these neuroimaging studies bolster the glutamate hypofunction hypothesis of schizophrenia and encourage further development of glutamatergic-based treatments that increase the activity of NMDAR [23].

Unfortunately, despite significant efforts, most PET tracers have shown suboptimal in vivo performance, including low brain uptake, rapid metabolism, and nonspecific binding. To date, no radioligands have been conclusively validated as in vivo imaging agents for NMDAR by PET or SPECT imaging [37, 38]. In addition, limited progress has been achieved in the development of radiotracers for kainate receptors, and no subtype-selective tracer has yet been reported. However, our group has successfully developed a novel PET tracer for AMPA receptors. This tracer, a derivative of 4-[2-(phenylsulfonylamino)ethylthio]-2,6-difluoro-phenoxyacetamide radiolabeled with ^{11}C ($[^{11}C]K$-2), showed specific binding to AMPA receptors [39].

In terms of GABA receptor imaging, there were insufficient PET/SPECT receptor availability studies for meta-analyses, though several reports exist.

Marques et al. reported a PET radiotracer $[^{11}C]Ro15$–4513 that specifically binds to the α5 subtype of the GABAA receptor (α5-GABA$_A$Rs), and antipsychotic-free patients with schizophrenia have lower α5-GABA$_A$Rs levels in the hippocampus [40]. Another research study utilized the GABA$_A$ radioligand $[^{18}F]$ fluorofluma-zenil, revealing reduced binding in the right caudate nucleus among individuals at high risk of schizophrenia [41]. Additionally, $[^{11}C]$flumazenil, combined with a GABA membrane transporter blocker, indicated impaired GABA transmission in a group of unmedicated patients [42].

2.1.1 Glutamate

To date, several meta-analyses have been conducted on glutamatergic neurometabolites using ^{1}H-MRS. Marsman et al. conducted the first meta-analysis, including 28 studies with 647 patients with SCZ and 608 healthy controls (HC). They reported that levels of glutamatergic neurometabolites, specifically Glu levels in the medial frontal region, were decreased in patients with SCZ compared to HC [43]. Merritt et al. performed another meta-analysis of ^{1}H-MRS studies in schizophrenia, finding significant elevations in Glu and Glx in the basal ganglia (BG) and Gln in the thalamus, as well as Glx in the medial temporal lobe. However, they reported no significant changes in Glu, Gln, or Glx levels in the mPFC of patients with schizophrenia [44]. Although no connections were observed in cortical areas, these results indicate that schizophrenia may involve increased levels of neurometabolites related to glutamate in various brain regions, supporting the hypothesis of heightened glutamatergic neurotransmission in the disorder [44].

Furthermore, a recent mega-analysis, including 59 studies with 1686 patients and 1451 HCs, reported using ^{1}H-MRS that glutamatergic neurometabolite levels in the mPFC and the anterior cingulate cortex (ACC) were lower in patients with SCZ compared to HCs, and these levels were associated with the severity of symptoms [45]. Given that the glutamatergic system is widely distributed to the whole brain's neural activity, it holds promise for delineating classifications based on treatment resistance in schizophrenia, stages of the disorder, and the nature of localized activity. Several studies, particularly focusing on individuals at high risk for psychosis (HR), have found significantly lower levels of Glu in the thalamus [46]. Additionally, another meta-analysis on antipsychotic-naïve or free patients with schizophrenia revealed no changes in glutamate levels using ^{1}H-MRS [47]. The findings regarding treatment-resistant schizophrenia (TRS) are mixed. One meta-analysis identified significantly higher ACC Glu levels in TRS compared to non-TRS [48], while a more recent meta-analysis found no significant differences in Glx or Glu levels in the ACC and dorsal striatum across the groups (HC, TRS, non-TRS, URS) [49]. Furthermore, one study reported that patients with TRS who did not respond to clozapine (ultra-resistant schizophrenia, URS) exhibited higher glutamatergic neurometabolite levels in the mPFC and the ACC in comparison with HCs, unlike non-TRS patients [50].

Recently, a larger meta-analysis that included 134 studies with 7993 participants with schizophrenia-spectrum disorders and 8744 HCs focusing on both glutamatergic and GABAergic neurometabolite with ^{1}H-MRS was reported [51]. This groundbreaking meta-analysis is the first in this field that enables to distinguish detailed regions of the brain and conduct subgroup analyses based on clinical stages and treatment responses. In their analysis of various subgroups, the researchers discovered across all groups— whether categorized by their stage of illness or status of treatment resistance such as first episode psychosis (FEP), unmedicated SCZ, and TRS—the levels of Glx in the basal ganglia (BG) were consistently elevated. This suggests that increased glutamatergic neurometabolite levels in the BG could be considered a characteristic trait of schizophrenia. Additionally, subgroup analyses revealed elevated Glx levels in the dorsolateral prefrontal cortex (DLPFC) and hippocampus, while other groups showed no differences, suggesting that antipsychotics may normalize elevated Glx levels of the hippocampus [51]. The administration of ketamine is thought to induce the hypofunction of parvalbumin interneurons, leading to an increase in Glx levels in the hippocampus of HC [35, 51]. Moreover, studies involving individuals with schizophrenia have observed decreased NMDAR availability in the hippocampus of drug-free patients. Collectively, this body of neuroimaging research supports the glutamate hypofunction hypothesis of schizophrenia and encourage further development of glutamatergic-based treatments that increase the activity of NMDAR [23].

2.1.2   GABA

IFrom the recent meta-analysis [51], compared to 84 studies focusing on Glx, 68 on Glu, and 11 on Gln, only 30 studies addressed GABA. This meta-analysis revealed that GABA levels in the MCC were decreased in the FEP group, the unmedicated group, and patient group compared with HCs. Therefore, GABA levels in the MCC may decrease with the onset of psychosis. In the occipital cortex, patients with schizophrenia group had lower GABA levels than HCs, a reduction that may be linked to the pathophysiology of schizophrenia-spectrum disorders [51]. In the frontal cortex, frontal white matter, occipito-parietal cortex, temporal cortex, and cerebellum, significant differences were found in neither the main meta-analyses nor subgroup analyses. The meta-analysis could not be performed for the DLPFC, temporal white matter, and hippocampus due to insufficient data.

Despite the hippocampus's critical role in excitatory/inhibitory (E/I) balance, data accumulation regarding GABA has been limited. A recent 7 T MRS study found no significant differences in GABA, Glu, Gln and Gln/Glu ratio. However, it identified lower DLPFC Glu and lower hippocampal GABA in individuals with a longer illness duration (more than 5 years) compared to those with a shorter duration [52]. Despite the wealth of data available, a critical need remains for more detailed information on key regions involved in glutamatergic and GABAergic activity to further understand the relationship between glutamate and schizophrenia. Nevertheless, MRS remains the most reliable modality for this research, offering insights that are often more dependable than provided by other imaging techniques like PET and SPECT.

2.2   Genetics

Large genome-wide association studies (GWAS), involving 36,989 Schizophrenia cases and 113,075 controls have identified 108 loci conservatively associated with schizophrenia. It was noted that several genes linked to this disorder encode proteins integral to glutamatergic neurotransmission. These include the mGlu3R (GRM3), the NR2 subunit of the NMDA receptor (GRIN2A), the serine racemase (SRR) and the AR1 subunit of the AMPA receptor (GRIA1). Furthermore, the studies discovered an overlap between genes affected by rare variants in schizophrenia and those identified within GWAS loci, showing broad convergence in the functions of some of the clusters of genes implicated by both sets of genetic variants. This is particularly true for genes related to abnormal glutamatergic synaptic function and calcium channel function [53].

2.3 Neurocircuits   Combining the findings from ^{1}H-MRS regarding the glutamatergic system with other modalities such as genetic, postmortem, other neuroimaging techniques such as PET, SPECT, functional MRI (fMRI), Diffusion Tensor Imaging (DTI), and animal models, is beneficial for elucidating the neuronal network. This approach aids in categorizing schizophrenia and identifying potential treatment targets.

Given that glutamate activity observed in ^{1}H-MRS each locus— mPFC, ACC, DLPFC—shows altered signals, the relationship between these regions is crucial for a deeper understanding of the neuronal networks involved in schizophrenia.

Recently, Wada et al. reported a review about possible key neurobiological basis of schizophrenia by integrating ^{1}H-MRS findings and animal studies [54]. The ventral tegmental area (VTA) is a region of the rat brain that is analogous to the substantia nigra (SN) in humans and is the origin of dopaminergic cell bodies. Some preclinical studies have reported that VTA receives projections from the mPFC via ventral subiculum of the hippocampus [55–59]. Some animal studies have shown that NMDAR blockade, including with substances like ketamine, leads to increased extracellular glutamatergic levels and heightened activity in the rat ventral subiculum of the hippocampus [56, 57]. This hyperactivity correlates with a rise in the number of spontaneously active dopamine neurons in the VTA [58]. Zimmerman et al. demonstrated that the infralimbic subdivision of the mPFC can significantly regulate the drive of the dopaminergic neurons from the VTA through the nucleus reuniens of the midline thalamus by activating the ventral subiculum of the hippocampus [59]. Similarly, Floresco et al. found that stimulating the ventral subiculum of the hippocampus activates glutamatergic afferents to the nucleus accumbens (NAc). The GABAergic projections from the NAc then inhibit neurons in the ventral pallidum, lifting tonic inhibition off the dopaminergic neurons in the VTA [58]. Integrating these rodent model findings, the glutamatergic dysfunction in the mPFC or hippocampus may be responsible for the increased potential for dopamine synthesis along the nigrostriatal pathway in schizophrenia [54]. Furthermore, MAM rodent model of schizophrenia has also revealed epigenetically-mediated alterations in NMDAR subunit composition [60], where the NMDAR dysfunction appeared earlier than dopamine hyper-synthesis [61]. Moreover, these studies indicate that NMDAR hypofunction and the resulting abnormal functioning in various brain regions, such as hyperactivity in the human anterior hippocampus, are core features of schizophrenia. Additionally, these findings suggest that glutamatergic dysfunction may be located temporally and spatially upstream of dopaminergic dysfunction in patients with schizophrenia and this hypothesis may serve as a complement to the limitations of the dopamine hypothesis [54].

2.4 Treatment

Numerous compounds targeting glutamate have been evaluated in clinical trials to date. NMDA receptor co-agonists, including D-cycloserine, D-serine, and glycine, were among the earliest to be researched in an attempt to enhance the functioning of NMDA-mediated interneurons [62].

There are several future drug targets related to the glutamatergic system. Gomes and Grace reported a review about new targets for schizophrenia, with a focus on glutamatergic system. An advanced understanding of the circuitry of schizophrenia has pointed to pathological origins in the excitation/inhibition (E/I) balance, particularly in the hippocampus, could be central to the disorder.

Compensating for the loss of parvalbumin (PV) interneurons, which is implicated in causing hippocampal hyperactivity may offer a more effective approach to alleviate a broad spectrum of schizophrenia symptoms [63]. This is underscored by evidence that hippocampal hyperactivity due to the loss of PV interneurons may contribute to the hyperdopaminergic state associated with schizophrenia [64, 65].

This evidence, alongside other studies indicating NMDA receptor hypofunction in schizophrenia, has bolstered efforts to develop drugs that can enhance NMDA receptor activity without causing excitotoxicity.

Glycine and D-serine, co-agonists of the NMDAR: Selective glycine transporter 1 (GlyT1) inhibitors such as sarcosine and bitopertin, have also been tested as an alternative method to increase the availability of glycine at NMDA receptors [66]. These compounds have shown some beneficial effects on positive, negative, and cognitive symptoms in schizophrenia [62, 67–71]; however, these effects were not consistently replicated in larger studies [72, 73]. Moreover, while the use of these compounds may be limited by tolerability issues, they also have the potential to enhance NMDA receptor function in schizophrenia [74]. Two meta-analyses found that glycine, when used adjunctively with non-clozapine antipsychotics, improves multiple symptom domains; conversely, symptoms worsened when glycine was added to clozapine [71, 75].

D-cycloserine, known as a partial agonist of NMDARs, has been studied for its potential benefits in schizophrenia. Additionally, the inhibition of D-amino acid oxidase (DAAO) with compounds such as sodium benzoate has recently been explored as a method of enhancing NMDA receptor activation by blocking the metabolism of D-amino acids. Sodium benzoate has shown promising results as an adjunctive therapy in early clinical trials [76–78].

Metabotropic glutamate receptor 2/3 (mGluR2/3) agonists are known to inhibit glutamate release from presynaptic terminals. One such agonist, pomaglumetad, developed by Eli Lilly, was shown to reduce hippocampal hyperactivity in the MAM model, leading to the downstream normalization of dopamine neuron activity in the VTA [65]. In patients with schizophrenia, pomaglumetad demonstrated beneficial effects on both positive and negative symptoms as a monotherapy in early phase 2 clinical trials [79]. However, it failed to show efficacy as a monotherapy or adjunct therapy in subsequent trials [80–83]. Later analyses of trial data suggested that certain subpopulations, particularly those in the earlier phases of illness or based on medication exposure during the two years before study entry, might respond better to pomaglumetad [84].

GABA: Alpha-5 $GABA_A$:($\alpha5GABA_A$) receptor positive allosteric modulator [85]: The $\alpha5GABA_A$ receptor is anticipated to serve as a promising target for antipsychotic treatments aimed at stabilizing hippocampal activity [86]. A positive allosteric modulator (PAM) that targets at $\alpha5GABA_A$ receptors has been effectively tested in the MAM model, lowering the activity of dopamine neuron populations and reducing amphetamine-triggered excessive movement to normal levels. This outcome was observed whether the PAM was administered system-wide or directly into the ventral hippocampus [87].

3 Conclusion and Future Direction

Research into schizophrenia has long been grounded in the dopamine hypothesis. To date, most antipsychotic medications still follow the same principles established in the 1950s. Many patients continue to suffer from poor response to treatment. The journey searching for new treatment targets has begun from decades ago, and the most promising hypothesis are now shifting from dopamine hypothesis to glutamatergic hypothesis, including the NMDAR hypofunction hypothesis and E/I imbalance hypothesis.

Unfortunately, promising glutamatergic-based compounds such as bitopertin and pomaglumetad, which had reached phase 3 development in clinical trials, have failed. Despite these setbacks, the glutamatergic system—along with the GABAergic system or muscarinic system, which we haven't discussed in this chapter—remains a highly promising area for future schizophrenia research. We continue to believe that exploring these systems will unlock new avenues for understanding and treating schizophrenia.

However, we continue to believe that exploring neurotransmitter systems, such as the glutamatergic, GABAergic system, and muscarinic system represents the most promising approach to unraveling the complexities of schizophrenia. These systems are likely to be pivotal in the future of schizophrenia research.

Despite the availability of many neuroimaging modalities to date, ^{1}H-MRS stands out one of the most reliable methods for assessing glutamatergic function. Its reliability especially notable given that radioligands for NMDA and AMPA receptors are yet to become widely used in this field.

As a future direction, we need to accumulate more data on local brain glutamate and GABA levels using ^{1}H-MRS, especially from key neurocircuit regions such as the hippocampus, DLPFC, ACC. Additionally, conducting subgroup analyses based on disease stage (at-risk mental state (ARMS), first-episode psychosis (FEP), chronic schizophrenia (SCZ), etc.), previous medication history (antipsychotic-free or naïve, clozapine users (treatment-resistant schizophrenia, TRS), clozapine nonresponders (ultra-resistant schizophrenia, URS), non-TRS, etc.), genetics, or specific pathophysiology will be crucial as a future research anker point.

Acknowledgments

I sincerely appreciate Dr. Masakata Wada for his valuable discussions and insights before the initiation of this manuscript. I am also grateful to Dr. Shinichiro Nakajima for providing me with the opportunity to write this chapter and for reviewing the final draft. Additionally, I would like to express my deepest gratitude to Dr. Hiroyuki Uchida, whose mentorship and guidance have greatly influenced my academic and professional growth.

References

1. Bruijnzeel D, Tandon R (2011) The concept of schizophrenia: From the 1850s to the DSM-5. Psychiatr Ann 41(5):289–295

2. Saha S, Chant D, Welham J, McGrath J (2005) A systematic review of the prevalence of schizophrenia. PLoS Med2(5):e141

3. McGrath J, Saha S, Chant D, Welham J (2008) Schizophrenia: a concise overview of incidence, prevalence, and mortality. Epidemiol Rev 30(1):67–76

4. Chang WC, Wong CSM, Chen EYH et al (2017) Lifetime prevalence and correlates of schizophrenia-spectrum, affective, and other non-affective psychotic disorders in the Chinese Adult Population. Schizophr Bull 43(6): 1280–1290

5. Hjorthoj C, Stürup AE, McGrath JJ et al (2017) Years of potential life lost and life expectancy inschizophrenia: a systematic review and meta-analysis. Lancet Psychiatry 4(4):295–301. https://doi.org/10.1016/S2215-0366(17)30078-0

6. Barbui C, Cipriani A (2008) Cognitive improvements with antipsychotics: real or practice effect? Evid Based Ment Health 11(2):42

7. Polese D, Fornaro M, Palermo M, De Luca V, De Bartolomeis A (2019) Treatment-resistant to antipsychotics: a resistance to everything? Psychotherapy in treatment-resistant schizophrenia and nonaffective psychosis: a 25-year systematic review and exploratory meta-analysis. Front Psych 10:210. https://doi.org/10.3389/fpsyt.2019.00210

8. Elkis H, Buckley PF (2016) Treatment-resistant schizophrenia. Psychiatr Clin North Am 39(2):239–265. https://doi.org/10.1016/j.psc.2016.01.006

9. Buckley P, Miller A, Olsen J, Garver D, Miller DD, Csernansky J (2001) When symptoms persist: clozapine augmentation strategies. Schizophr Bull 27(4):615–628. https://doi.org/10.1093/oxfordjournals.schbul.a006901

10. Siskind D, Siskind V, Kisely S (2017) Clozapine response rates among people with treatment-resistant schizophrenia: data from a systematic review and meta-analysis. Can J Psychiatr 62(11):772–777. https://doi.org/10.1177/0706743717718167

11. Kane JM, Correll CU (2016) The role of clozapine in treatment-resistant schizophrenia.

JAMA Psychiatry 73(3):187. https://doi.org/10.1001/jamapsychiatry.2015.2966

12. Kim JS, Kornhuber HH, Schmid-Burgk W, Holzmüller B (1980) Low cerebrospinal fluid glutamate in schizophrenic patients and a new hypothesis on schizophrenia. Neurosci Lett 20: 379–382

13. Rothman DL, Behar KL, Hyder F, Shulman RG (2003) In Vivo NMR Studies of the Glutamate Neurotransmitter Flux and Neuroenergetics: Implications for Brain Function. Annu Rev Physiol 65:401–427

14. Reiner A, Levitz J (2018) Glutamatergic signaling in the central nervous system:ionotropic and metabotropic receptors in concert. Neuron 98:1080–1098. https://doi.org/10.1016/j.neuron.2018.05.018

15. Li CT, Yang KC, Lin WC (2019) Glutamatergic dysfunction and glutamatergic compounds for major psychiatric disorders: evidence from clinical neuroimaging studies. Front Psych 10: 767–767. https://doi.org/10.3389/fpsyt.2018.00767

16. Javitt DC, Zukin SR (1991) Recent advances in the phencyclidine model of schizophrenia. Am J Psychiatry 148:1301–1308

17. Olney JW, Farber NB (1995) Glutamate receptor dysfunction and schizophrenia. Arch Gen Psychiatry 52:998–1007

18. Dalmau J, Gleichman AJ, Hughes EG, Rossi JE, Peng X, Lai M, Dessain SK, Rosenfeld MR, Rita Balice-Gordon R, Lynch DR (2008) Anti-NMDA-receptor encephalitis: case series and analysis of the effects of antibodies. Lancet Neurol 7(12):1091–1098. https://doi.org/10.1016/S1474-4422(08)70224-2

19. Dalmau J, Tuzun E, Wu HY et al (2007) Paraneoplastic anti-N-methyl-D-aspartate receptor encephalitis associated with ovarian teratoma. Ann Neurol 61(1):25–36. https://doi.org/10.1002/ana.21050

20. Krystal JH, Perry EB Jr, Gueorguieva R et al (2005) Comparative and interactive human psychopharmacologic effects of ketamine and amphetamine: implications for glutamatergic and dopaminergic model psychoses and cognitive function. Arch Gen Psychiatry 62(9): 985–994

21. Javitt DC (2010) Glutamatergic theories of schizophrenia. Isr J Psychiatry Relat Sci 47(1): 4–16

22. Perez SM, Lodge DJ (2013) Hippocampal interneuron transplants reverse aberrant dopamine system function and behavior in a rodent model of schizophrenia. Mol Psychiatry 18: 1193–1198

23. Poels EM, Kegeles LS, Kantrowitz JT et al (2014) Imaging glutamate in schizophrenia: review of findings and implications for drug discovery. Mol Psychiatry 19(1):20–29. https://doi.org/10.1038/mp.2013.136

24. Kreis R, Ernst T, Ross BD (1993) Development of the human brain: in vivo quantification of metabolite and water content with proton magnetic resonance spectroscopy. Magn Reson Med 30(4):424–437. https://doi.org/10.1002/mrm.1910300405

25. Bertolino A, Weinberger DR (1999) Proton magnetic resonance spectroscopy in schizophrenia. Eur J Radiol 30(2):132–141

26. Dager SR, Corrigan NM, Richards TL, Posse S (2008) Research applications of magnetic resonance spectroscopy to investigate psychiatric disorders. Top Magn Reson Imaging 19(2): 81–96

27. Snyder J, Wilman A (2010) Field strength dependence of PRESS timings for simultaneous detection of glutamate and glutamine from 1.5 to 7T. J MagnReson 203(1):66–72. https://doi.org/10.1016/j.jmr.2009.12.002

28. Gillies RJ (1992) Nuclear magnetic resonance and its applications to physiological problems. Annu Rev Physiol 54(1):733–748

29. Rothman DL, Sibson NR, Hyder F, Shen J, Behar KL, Shulman RG (1999) In vivo nuclear magnetic resonance spectroscopy studies of the relationship between the glutamate glutamine neurotransmitter cycle and functional neuro energetics. Philos Trans R Soc Lond B Biol Sci 354(1387):1165–1177

30. Jackson ME, Homayoun H, Moghaddam B (2004) NMDA receptor hypofunction produces concomitant firing rate potentiation and burst activity reduction in the prefrontal cortex. Proc Natl Acad Sci–PNAS 101(22): 8467–8472

31. Lahti AC, Weiler MA, Tamara Michaelidis BA et al (2001) Effects of ketamine in normal and schizophrenic volunteers. Neuropsychopharmacology 25(4):455–467

32. Kraguljac NV, CarleM FMA, Tran S, Yassa MA, White DM et al (2018) Mnemonic Discrimination Deficits in First-Episode Psychosis and a Ketamine Model Suggests Dentate Gyrus Pathology Linked to N-Methyl-D-Aspartate Receptor Hypofunction. Biol Psychiatry: Cognit Neurosci Neuroimaging 3(3):231–238

33. Krystal JH, Karper LP, Seibyl JP et al (1994) Subanesthetic effects of the noncompetitive NMDA antagonist, ketamine, in humans: psychotomimetic, perceptual, cognitive, and neuroendocrine responses. Arch Gen Psychiatry 51(3):199–214

34. Rowland LM, Bustillo JR, Mullins PG et al (2005) Effects of ketamine on anterior cingulate glutamate metabolism in healthy humans: a 4-T proton MRS study. Am J Psychiatry 162(2):394–396

35. Kraguljac NV, Frölich MA, Tran S, White DM, Nichols N, Barton-Mcardle A et al (2017) Ketamine modulates hippocampal neurochemistry and functional connectivity: A combined magnetic resonance spectroscopy and resting-state fMRI study in healthy volunteers. Mol Psychiatry 22(4):562–569

36. Stone JM, Dietrich C, Edden R, Mehta MA, De Simoni S, Reed LJ et al (2012) Ketamine effects on brain GABA and glutamate levels with 1H-MRS: relationship to ketamine-induced psychopathology. Mol Psychiatry 17(7):664–665

37. van der Aart J, Golla SSV, van der Pluijm M, Schwarte LA, Schuit RC, Klein PJ et al (2018) First in human evaluation of [18F]PK-209, a PET ligand for the ion channel binding site of NMDA receptors. EJNMMI Res 8(1):69–12. https://doi.org/10.1186/s13550-018-0424-2

38. Dierckx RAJO, Otte A, de Vries EFJ, van Waarde A, Lammertsma AA (2021) PET and SPECT of Neurobiological Systems, 2nd edn. Springer, Cham. https://doi.org/10.1007/978-3-030-53176-8

39. Miyazaki T, Nakajima W, Hatano M, Shibata Y, Kuroki Y, Arisawa T, Serizawa A et al (2020) Visualization of AMPA receptors in living human brain with positron emission tomography. Nat Med 26(2):281–288. https://doi.org/10.1038/s41591-019-0723-9

40. Marques TR, Ashok AH, Angelescu I, Borgan F, Myers J, Lingford-Hughes A, Nutt DJ, Mattia Veronese M, Turkheimer FE, Howes OD (2021) GABA-A receptor differences in schizophrenia: a positron emission tomography study using [11C] Ro154513. Mol Psychiatry 26(6):2616–2625. https://doi.org/10.1038/s41380-020-0711-y

41. Kambeitz J, Abi-Dargham A, Kapur S, Howes OD (2014) Alterations in cortical and extra-striatal subcortical dopamine function in schizophrenia: systematic review and meta-analysis of imaging studies. Br J Psychiatry 204(6):420–429

42. Frankle WG, Cho RY, Prasad KM, Mason NS, Paris J, Himes ML et al (2015) In vivo measurement of GABA transmission in healthy subjects and schizophrenia patients. Am J Psychiatry 172(11):1148–1159

43. Marsman A, van den Heuvel MP, Klomp DWJ, Kahn RS, Luijten PR, Hulshoff HE et al (2013) Glutamate in schizophrenia: a focused review and meta-analysis of 1H-MRS studies. Schizophr Bull 39(1):120–129

44. Merritt K, Egerton A, Kempton MJ et al (2016) Nature of glutamate alterations in schizophrenia: a meta-analysis of proton magnetic resonance spectroscopy studies. JAMA Psychiat 73(7):665–674. https://doi.org/10.1001/jamapsychiatry.2016.0442

45. Merritt K, McGuire PK, Egerton A, Aleman A, Block W, Bloemen OJN et al (2021) 1H-MRS in Schizophrenia Investigators, Aleman A, Block W, et al. Association of age, antipsychotic medication, and symptom severity in schizophrenia with proton magnetic resonance spectroscopy brain glutamate level: a mega-analysis of individual participant-level data. JAMA Psychiat (Chicago, III) 78(6):667–668. https://doi.org/10.1001/jamapsychiatry.2021.0380

46. Wenneberg C, Glenthøj BY, Hjorthøj C, Zingenberg FJB, Glenthøj LB, Rostrup E et al (2020) Cerebral glutamate and GABA levels in high-risk of psychosis states: a focused review and meta-analysis of 1H-MRS studies. Schizophrenia Res 215:38–48

47. Iwata Y, Nakajima S, Plitman E, Mihashi Y, Caravaggio F, Chung JK et al (2018) Neurometabolite levels in antipsychotic-naive/free patients with schizophrenia: a systematic review and meta-analysis of H-1-MRS studies. Prog Neuro-Psychopharmacol Biol Psychiatry 86:340–352. https://doi.org/10.1016/j.pnpbp.2018.03.016

48. Kumar V, Manchegowda S, Jacob A, Rao NP (2020) Glutamate metabolites in treatment resistant schizophrenia: A meta-analysis and systematic review of H-1-MRS studies. Psychiatry Res Neuroimaging 300:111080. https://doi.org/10.1016/j.pscychresns.2020.111080

49. Smucny J, Carter CS, Maddock RJ (2023) Greater choline-containing compounds and myo-inositol in treatment-resistant versus responsive schizophrenia: a 1H-magnetic resonance spectroscopy meta-analysis. Biol Psychiatry: Cognit Neurosci Neuroimaging 9(2):137–145. https://doi.org/10.1016/j.bpsc.2023.10.008

50. Iwata Y, Nakajima S, Plitman E, Caravaggio F, Kim J, Shah P et al (2019) Glutamatergic Neurometabolite Levels in Patients With Ultra-Treatment-Resistant Schizophrenia: A Cross-Sectional 3T Proton Magnetic Resonance Spectroscopy Study. Biol Psychiatry 85(7):596–605. https://doi.org/10.1016/j.biopsych.2018.09.009

51. Nakahara T, Tsugawa S, Noda Y, Ueno F, Honda S, Kinjo M et al (2022) Glutamatergic and GABAergic metabolite levels in schizophrenia-spectrum disorders: a meta-

analysis of 1H-magnetic resonance spectroscopy studies. Mol Psychiatry 27(1):744–757. https://doi.org/10.1038/s41380-021-01297-6

52. Wijtenburg SA, Wang M, Korenic SA, Chen S, Barker PB, Rowland LM (2021) Metabolite alterations in adults with schizophrenia, first degree relatives, and healthy controls: a multiregion 7T MRS study. Front Psych 12:656459–656459. https://doi.org/10.3389/fpsyt.2021.656459

53. Consortium SWG of TPG, Schizophrenia Working Group of the Psychiatric Genomics Consortium (2014) Biological insights from 108 schizophrenia-associated genetic loci. Nature 511:421–427

54. Wada M, Noda Y, Iwata Y, Tsugawa S, Yoshida K, Tani H et al (2022) Dopaminergic dysfunction and excitatory/inhibitory imbalance in treatment-resistant schizophrenia and novel neuromodulatory treatment. Mol Psychiatry 27(7):2950–2967. https://doi.org/10.1038/s41380-022-01572-0

55. Patton MH, Bizup BT, Grace AA (2013) The infralimbic cortex bidirectionally modulates mesolimbic dopamine neuron activity via distinct neural pathways. J Neurosci 33(43):16865–16873

56. Homayoun H, Moghaddam B (2007) NMDA receptor hypofunction produces opposite effects on prefrontal cortex interneurons and pyramidal neurons. J Neurosci 27(43):11496–11500. https://doi.org/10.1523/JNEUROSCI.2213-07.2007

57. Schobel SA, Chaudhury NH, Khan UA, Paniagua B, Styner MA, Asllani I et al (2013) Imaging patients with psychosis and a mouse model establishes a spreading pattern of hippocampal dysfunction and implicates glutamate as a driver. Neuron (Cambridge, Mass.) 78(1):81–93

58. Floresco SB, Todd CL, Grace AA (2001) Glutamatergic afferents from the hippocampus to the nucleus accumbens regulate activity of ventral tegmental area dopamine neurons. J Neurosci 21(13):4915–4922

59. Zimmerman EC, Grace AA (2016) The nucleus reuniens of the midline thalamus gates prefrontal-hippocampal modulation of ventral tegmental area dopamine neuron activity. J Neurosci 36(34):8977–8984

60. Gulchina Y, Xu S, Snyder MA, Elefant F, Gao W (2017) Epigenetic mechanisms underlying NMDA receptor hypofunction in the prefrontal cortex of juvenile animals in the MAM model for schizophrenia. J Neurochem 143(3):320–333

61. Flagstad P, Mørk A, Glenthøj BY, van Beek J, Michael-Titus AT, Didriksen M (2004) Disruption of neurogenesis on gestational day 17 in the rat causes behavioral changes relevant to positive and negative schizophrenia symptoms and alters amphetamine-induced dopamine release in nucleus accumbens. Neuropsychopharmacology 29(11):2052–2064

62. Tsai G, Yang P, Chung LC, Lange N, Coyle JT (1998) D-serine added to antipsychotics for the treatment of schizophrenia. Biol Psychiatry 44(11):1081–1089

63. Gomes FV, Grace AA (2021) Beyond dopamine receptor antagonism: new targets for schizophrenia treatment and prevention. Int J Mol Sci 22(9):4467. https://doi.org/10.3390/ijms22094467

64. Grace AA, Gomes FV (2019) The circuitry of dopamine system regulation and its disruption in schizophrenia: insights into treatment and prevention. Schizophr Bull 45(1):148–157

65. Sonnenschein SF, Gomes FV, Grace AA (2020) Dysregulation of midbrain dopamine system and the pathophysiology of schizophrenia. Front Psych 11:613–613

66. Tsai G, Lane HY, Yang P, Chong MY, Lange N (2004) Glycine transporter I inhibitor, N-methylglycine (sarcosine), added to antipsychotics for the treatment of schizophrenia. Biol Psychiatry 55(5):452–456

67. Heresco-Levy U, Ermilov M, Lichtenberg P, Bar G, Javitt DC (2004) High-dose glycine added to olanzapine and risperidone for the treatment of schizophrenia. Biol Psychiatry 55(2):165–171

68. Heresco-Levy U, Javitt DC, Ermilov M, Mordel C, Silipo G, Lichtenstein M (1999) Efficacy of high-dose glycine in the treatment of enduring negative symptoms of schizophrenia. Arch Gen Psychiatry 56(1):29–36

69. Heresco-Levy U, Javitt DC, Ebstein R, Vass A, Lichtenberg P, Bar G et al (2005) D-serine efficacy as add-on pharmacotherapy to risperidone and olanzapine for treatment-refractory schizophrenia. Biol Psychiatry 57(6):577–585

70. Kantrowitz JT, Malhotra AK, Cornblatt B, Silipo G, Balla A, Suckow RF et al (2010) High dose D-serine in the treatment of schizophrenia. Schizophr Res 121(1):125–130

71. Tsai GE, Lin P-Y (2010) Strategies to enhance N-methyl-D-aspartate receptor-mediated neurotransmission in schizophrenia, a critical review and meta-analysis. Curr Pharm Des 16(5):522–537

72. Buchanan RW, Javitt DC, Marder SR, Schooler NR, Gold JM, McMahon RP et al (2007) The

Cognitive and Negative Symptoms in Schizophrenia Trial (CONSIST): The efficacy of glutamatergic agents for negative symptomsand cognitive impairments. Am J Psychiatry 164(10):1593–1602

73. Weiser M, Heresco-Levy U, Davidson M, Javitt DC, Werbeloff N, Gershon AA et al (2012) A multicenter, add-on randomized controlled trial of low-dose d-serine for negative and cognitive symptoms of schizophrenia. J Clin Psychiatry 73(6):e728–e734

74. Kantrowitz J, Javitt DC (2012) Glutamatergic transmission in schizophrenia: From basic research to clinical practice. Curr Opin Psychiatry 25(2):96–102

75. Singh SP, Singh V (2011) Meta-analysis of the efficacy of adjunctive NMDA receptor modulators in chronic schizophrenia. CNS Drugs 25(10):859–885. https://doi.org/10.2165/11586650-000000000-00000

76. Lin CH, Lin CH, Chang YC, Chen PW, Yang HT et al (2018) Sodium benzoate, a D-amino acid oxidase inhibitor, added to clozapine for the treatment of schizophrenia: a randomized, double-blind, placebo-controlled trial. Biol Psychiatry 84(6):422–432

77. Lin CY, Liang SY, Chang YC, Ting SY, Kao CL, Wu YH et al (2017) Adjunctive sarcosine plus benzoate improved cognitive function in chronic schizophrenia patients with constant clinical symptoms: a randomised, double-blind, placebo-controlled trial. World J Biol Psychiatry 18(5):357–368

78. Nicoletti F, Bockaert J, Collingridge GL, Conn PJ, Ferraguti F, Schoepp DD et al (2011) Metabotropic glutamate receptors: from the workbench to the bedside. Neuropharmacology 60(7):1017–1041

79. Martenyi F, Monn JA, Patil ST, Zhang L, Lowe SL, Jackson KA, Andreev BV, Avedisova AS, Bardenstein LM, Gurovich IY, Morozova MA et al (2007) Activation of mGlu2/3 receptors as a new approach to treat schizophrenia: a randomized Phase 2 clinical trial. Nat Med 13(9):1102–1107

80. Adams DH, Kinon BJ, Baygani S, Millen BA, Velona I, Kollack-Walker S, Walling DP (2013) A long-term, phase 2, multicenter, randomized, open-label, comparative safety study of pomaglumetad methionil (LY2140023 monohydrate) versus atypical antipsychotic standard of care in patients with schizophrenia. BMC Psychiatry 13(1):143–143. https://doi.org/10.1186/1471-244X-13-143

81. Stauffer VL, Millen BA, Andersen S, Kinon BJ, LaGrandeur L, Lindenmayer JP, Gomez JC et al (2013) Pomaglumetad methionil: no significant difference as an adjunctive treatment for patients with prominent negative symptoms of schizophrenia compared to placebo. Schizophr Res 150(2):434–441

82. Downing ACM, Kinon BJ, Millen BA, Zhang L, Liu L, Morozova MA et al (2014) A double-blind, placebo-controlled comparator study of LY2140023 monohydrate in patients with schizophrenia. BMC Psychiatry 14(1):351–351

83. Adams DH, Zhang L, Millen BA, Kinon BJ, Gomez JC (2014) Pomaglumetad methionil (LY2140023 Monohydrate) and aripiprazole in patients with schizophrenia: a phase 3, multicenter, double-blind comparison. Schizophr Res Treatment 2014(2014):758212–758211

84. Kinon BJ, Millen BA, Zhang L, McKinzie DL (2015) Exploratory analysis for a targeted patient population responsive to the metabotropic glutamate 2/3 receptor agonist pomaglumetad methionil in schizophrenia. Biol Psychiatry 78(11):754–762

85. Gill KM, Lodge DJ, Cook JM, Aras S, Grace AA (2011) A novel α5GABA(A)R-positive allosteric modulator reverses hyperactivation of the dopamine system in the MAM model of schizophrenia. Neuropsychopharmacology 36(9):1903–1911

86. Glykys J, Mann EO, Mody I (2008) Which GABAA receptor subunits are necessary for tonic inhibition in the hippocampus? J Neurosci 28(6):1421–1426

87. Biggs CS, Pearce BR, Fowler LJ et al (1992) The effect of sodium valproate on extracellular GABA and other amino acids in the rat ventral hippocampus: an in vivo micro dialysis study. Brain Res 594(1):138–142

Chapter 12

Artificial Intelligence for Molecular Brain Imaging

Jarrad Perron and Iman Beheshti

Abstract

In this work, we aim to give a thorough treatment of artificial intelligence in the domain of molecular neuroimaging. It will be divided between two broad portions. First, we will give a brief overview of the usage of molecular imaging methods within clinical practice, before presenting the principles and mechanisms of nuclear imaging. Second, we will discuss image preprocessing operations relevant to studies performed on neuroimaging data and give a thorough review of many methods within traditional machine learning and the more recent field of deep learning.

Key words Alzheimer's disease, Parkinson's disease, Dementia, Mild cognitive impairment, Biomarker, Machine learning, Deep learning, Positron emission tomography, Single-photon emission computed tomography

1 Introduction - Clinical Utility of Nuclear Imaging

Nuclear imaging methods are based on sub-pharmacological dosages of radioactive substances called tracers. These tracers are radioactive substances which decay and release photons either directly or as a by-product of their decay. These photons are detected and measured by specialized imaging methodologies called positron emission tomography (PET) and single-photon emission computed tomography (SPECT). Measuring the distribution of these tracers through their decay photons is a widespread and clinically useful means for noninvasive and in vivo characterization, measurement, and visualization of physical processes [1]. The ability of PET and SPECT imaging methods to capture dynamic physiological processes sets them apart from structural-anatomical approaches and gives them a place of prominence within diagnostic medical imaging.

PET and SPECT imaging have a broad range of clinical usages. We list here a representative survey of their important applications throughout some major fields of medicine and point the way toward an expansive literature on these topics.

Daichi Sone (ed.), *Molecular Imaging for Brain Diseases*, Neuromethods, vol. 222, https://doi.org/10.1007/978-1-0716-4494-2_12,
© The Author(s), under exclusive license to Springer Science+Business Media, LLC, part of Springer Nature 2025

PET has become an indispensable tool in oncology, transforming the clinical landscape of the practice for its ability to detect and characterize tumors. The workhorse of oncology PET is the glucose analog fluorodeoxyglucose (FDG). Since cancer cells are hypermetabolic to most surrounding tissue [2], this tracer shows hyperintensity on FDG-PET studies in oncology, which allows for the visualization of any abnormal cellular activity [3]. PET studies are also used to stage cancer progression by revealing the extent and spread of disease throughout the body and give information about the aggressiveness of tumors to guide treatment decisions [4, 5]. It is also useful in delineating tumor volumes during the planning of radiation therapy [4, 5]. PET imaging may be further used to monitor therapeutic response to make timely adjustments to treatment plans [3–5]. While PET may be the more common modality, SPECT imaging is also used for the study of neuroendocrine tumors and thyroid carcinoma [6] but is particularly valuable for its ability to detect osseous metastases, especially from prostate and breast cancers, and primary bone malignancies [6, 7].

Both PET and SPECT imaging have prominent roles in neurological practice and especially so in cognitive neurology. PET imaging is the standard of care for the differential diagnosis of neurodegenerative dementia disorders, where differences in spatial topography of brain metabolism and abnormal proteinopathy are signs of underlying pathology [8–14]. These may also be monitored by longitudinal nuclear neuroimaging, especially considering the recent approval of anti-amyloid drugs like aducanumab and lecanemab [15, 16]. SPECT imaging is specifically useful in differentiating and detecting movement-related neurodegenerative disorders such as Parkinson's disease and associated parkinsonian disorders such as dementia with Lewy bodies or multiple systems atrophy [17–20]. Brain SPECT is also well known for its use as a measure of cerebral blood flow. It is well suited to the detection of transient ischemic attacks, acute stroke, arterial occlusion, and subarachnoid hemorrhage [21]. SPECT may also be used for the localization of epileptic foci within the brain prior to surgical intervention [22]. In some cases, brain death may also be confirmed by perfusion imaging with SPECT [23, 24]. Pain medicine and research is another prominent application of nuclear neuroimaging, especially as a niche use of SPECT [25].

Nuclear imaging also has an ongoing and increasing literature in the diagnosis, detection, and monitoring of diseases with infectious or inflammatory origins, such as AIDS/HIV, tuberculosis, sarcoidosis, diabetic foot infections, and vasculitis [26].

The integration of SPECT and PET with computed tomography (SPECT-CT, PET-CT) or magnetic resonance (PET-MR) imaging has further expanded the capabilities of molecular neuroimaging by providing complementary anatomical and functional information within a single imaging session. While both SPECT-

CT and PET-CT have entered mainstream use and are considered a clinical norm, PET-MR imaging is yet to be widely distributed and is in the early adoption phase of the technology with uses limited to highly specialized imaging centers and research [27–31].

2 Materials - Preparatory Principles and Mechanisms of Nuclear Imaging

2.1 Basics of Radioactive Decay

The foundation principle upon which all nuclear imaging relies is that of radioactive decay. Radioactive decay is the process by which unstable nuclei stabilize through the emission of mass and energy [1].

2.1.1 Alpha Decay

In alpha decay, an unstable nucleus emits an alpha particle, composed of two protons and two neutrons, written as

$$\,^{A}_{Z}X \rightarrow \,^{A-4}_{Z-2}\Upsilon + \,^{4}_{2}\text{He}$$

where the alpha particle is synonymous with a helium-4 nucleus, $^{A}_{Z}X$ is the parent nucleus of atomic mass A and atomic number Z, and $^{A-4}_{Z-2}\Upsilon$ is the transmuted daughter nucleus [1].

The range of alpha particles is relatively short compared to other forms of radiation. Due to their size and charge, alpha particles experience strong interactions with the surrounding atoms, leading to rapid energy loss through ionization. The distance an alpha particle can travel in a material is known as its range. In air, for instance, alpha particles may only travel a few centimeters, while in denser materials like tissue or a thin sheet of paper, their range can be reduced to just a few micrometers. While alpha particles pose a lower external radiation hazard compared to other forms of radiation, the main concern arises when these substances are inhaled, ingested, or come into direct contact with sensitive tissues. Alpha-emitting isotopes can become hazardous sources if incorporated into the body, as the particles emitted close to living tissues can cause significant damage [1, 32, 33].

2.1.2 Beta Decay

In beta decay a neutron within the nucleus undergoes a transformation into a proton or vice versa, giving rise to the emission of a beta particle. Each of these transformations has unique decay equations, referred to respectively as beta-minus decay

$$n \rightarrow p + \beta^{-} + \bar{\nu}_{e}$$

where the transformation of the neutron releases an electron (β^{-}) and an electron anti-neutrino ($\bar{\nu}_{e}$), or as beta-plus (β^{+}) decay

$$p \rightarrow n + \beta^{+} + \bar{\nu}_{e}$$

where the reverse transformation releases a positron (β^{+}), also known as an anti-electron, and an electron neutrino ($\bar{\nu}_{e}$).

The range of beta particles is comparatively longer than alpha particles, owing to their smaller mass and charge. Beta particles can travel several centimeters in air and penetrate deeper into materials such as tissue. While beta particles pose an external radiation hazard more significant than with alpha decay, the primary concern arises when beta-emitting substances are introduced into the body through inhalation, ingestion, or direct contact [1, 32, 33].

A crucial distinction to make about beta-emitting substances in medical imaging is that neither the beta particles nor the neutrinos are ever directly measured. Rather, the positron emitted by beta-plus decay, being an antimatter particle, can annihilate a nearby electron in the nuclear neighborhood. This annihilation event results in the creation of two photons in the gamma ray range (511 keV). These annihilation photons are measured in nuclear imaging procedures using beta-emitting substances by tracing a line of response [1].

2.1.3 Gamma Decay

Gamma decay stands out as a process where an excited nucleus transitions to a lower energy state by emitting a high-energy photon called a gamma ray. Unlike alpha and beta decay, gamma decay does not alter the atomic or mass numbers of the nucleus but focuses solely on releasing excess energy from the parent nucleus, written as

$$_{Z}^{A}X^* \to {}_{Z}^{A}X + \gamma$$

where * denotes the excited state of the nucleus and γ signifies the emitted gamma ray.

The energy of the gamma photons is a critical parameter, quantified in electron volts (eV), determining their penetrative capacity and influencing their interaction with matter. The range of gamma rays is considerably greater than that of alpha and beta particles, owing to their nature as electromagnetic waves. Gamma rays can traverse long distances in air and penetrate various materials. While alpha particles may be stopped by thick protective equipment or glass and beta particles may be stopped by sheet of aluminum, gamma rays require several centimeters of lead to be effectively stopped [32, 34]. In common practice, lead-lined carriers, containers, bricks, and glasses are used for radiation protection.

2.2 Half-Life

Half-life, denoted as $t_{1/2}$ or τ, is the time required for half of a radioactive substance to undergo decay and transform into a more stable state. This temporal parameter is intrinsic to the inherent instability of radioactive isotopes and the probabilistic nature of decay processes. The longer the half-life, the more stable the isotope, and vice versa. This concept is central to the selection of isotopes for various applications, as it dictates the duration over which a given isotope remains active and contributes to imaging processes [1].

Regardless of the specific decay process, the time-average for N unstable nuclei to decay is given by

$$\frac{dN}{dt} = A = \lambda N$$

where A is the activity and λ is the decay constant unique to each radionuclide [1]. This differential equation has the solution

$$A(t) = A_0 \exp[-\lambda t]$$

where A_0 is the initial activity. The half-life of a substance is related to the decay constant by

$$t_{1/2} = \frac{\ln 2}{\lambda}.$$

2.3 Imaging Methods: SPECT and PET

When a patient is administered a radiopharmaceutical for a nuclear imaging procedure, gamma radiation is emitted by the decay or as a byproduct of that decay. Gamma cameras are designed to detect and record the gamma rays emitted by down-converting high-energy gamma photons to a lower energy before converting the optical signal to an electric one and amplifying it [35, 36].

Scintillation is the foundational principle of detector design. In an isolated atom we know there exist discrete allowed energy states, but in a crystal lattice there may exist a band of allowed energy states. The highest energy filled energy band is known as the valence band, and the lowest energy unfilled band is the conduction band. An electron within the valence band may absorb energy through either photoelectric interaction or Compton scattering before being excited into the conduction band and will eventually decay back to the valence band by emission of a photon. Most scintillation materials emit photons in the ultraviolet range but adding impurities to the crystal lattice structure can fine-tune the energy of the scintillation photons. Only materials with appropriate combinations of photon stopping power, scintillation decay time, light output, and energy resolution are appropriate for use in medical imaging [35].

The lower-energy photons are then passed through to a photo-multiplier tube (PMT). A PMT is effectively a series of dynodes capped with a cathode–anode pair. Incoming photons strike a photosensitive cathode which emits electrons upon perturbation. These ejected photons are accelerated through a series of dynodes and cause the ejection of an ever-increasing number of electrons. This electron avalanche eventually strikes an anode where this amplified current is output from the device [36]. Silicon photo-multipliers are a newer technology that may also be used for this purpose [35].

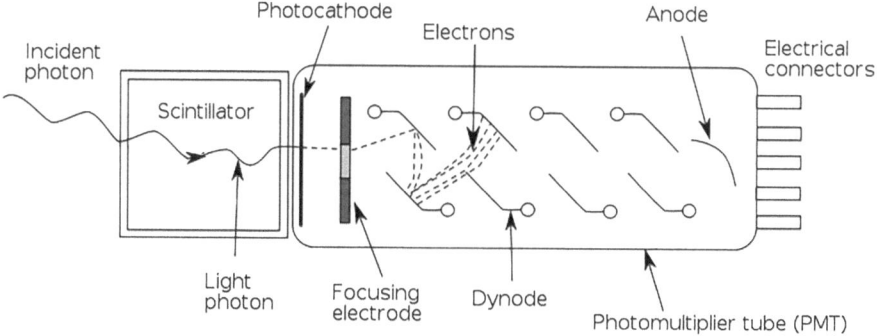

Fig. 1 Photomultiplier tube with scintillator and incident photon. Light is incoming from the left-hand side and causes scintillation within the crystal. These photons strike a photocathode which causes a small electric current. This current is amplified through the body of the PMT

In summary, when a gamma ray interacts with the crystal, it produces flashes of light, a phenomenon known as scintillation. Surrounding the crystal is a lead or tungsten shield called a collimator. This component ensures that only gamma rays traveling in specific directions reach the crystal, allowing for spatial localization of the radiation source. The PMT converts the scintillation flashes into electrical signals, creating a two-dimensional image of the gamma-ray distribution within the patient's body. *See* Fig. 1 for an illustration of this process.

SPECT devices are well described as one or more gamma cameras positioned strategically around a gantry that holds the patients. These cameras capture multiple projection images from various angles as they orbit around the patient (*see* Fig. 2). These 2D projections, representing the spatial distribution of gamma rays, form the foundation for the subsequent image reconstruction process—an algorithmic process by which 2D projections are transformed into 3D images of the spatial distribution of radiotracer within the patient [33, 37].

PET imaging is quite like SPECT; however, there are significant differences in key areas. In practice, PET devices are almost always a circularly arranged ring of detectors. PET imaging also uses beta-emitting substances over gamma-emitting as tracer. Rather than the direct emission of gamma rays, PET measures annihilation photons that result from the annihilation of positrons and electrons within the subject. These photons are emitted back-to-back due to conservation of momentum and will trace a line-of-response as they travel and strike detectors on opposing detectors across the ring. No physical collimator is employed in PET; instead, a computational approach called electronic collimation is employed. Data acquisition and event rejection in PET involve the simultaneous detection of paired gamma rays as a series of coincidence events (*see* Fig. 3). Unlike SPECT, where multiple projections are acquired, PET primarily focuses on these coincident

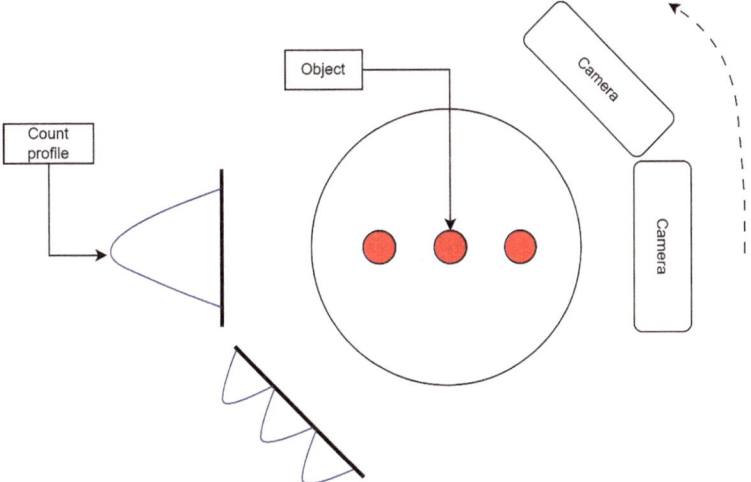

Fig. 2 Projection images with a single-head single-photon emission compute tomography (SPECT) camera. As the camera rotates about the objects to be imaged, each of the projections has different count intensities and distributions depending on the concentration and location of the radioactive material. These images are eventually processed into one complete image

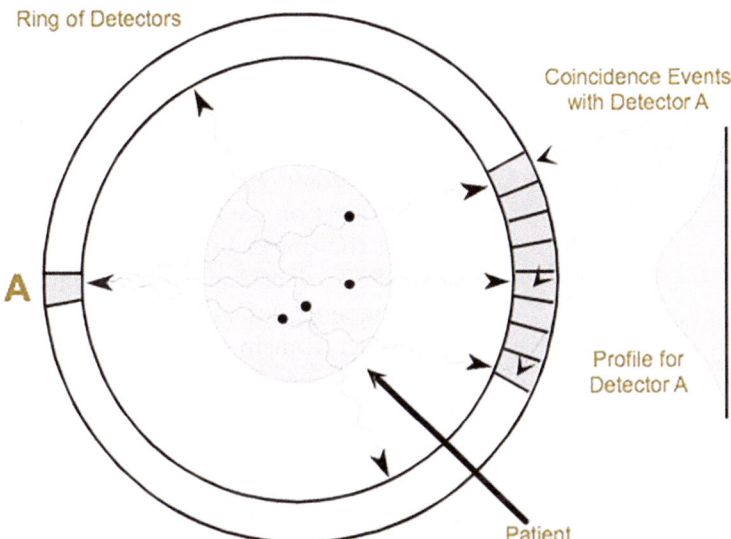

Fig. 3 Coincidence detection with a positron emission tomography (PET) device. Once a photon is registered at detector A on the left-hand side of the ring, the coincidence timing windows will open for all detectors within an arc on the opposite side. Note the several possible coincident photons on the right-hand side and the likelihood of these being the true coincident photons from the original annihilation event in the detector profile

events to a sinogram from which images may be reconstructed [1, 33, 36, 37].

Compared to one another, they both offer valuable means of imaging physiology and perfusion in vivo; however, there are some notable trade-offs [37]. PET imaging offers superior spatial resolution, especially in devices capable of time-of-flight acquisition, and PET is more useful in quantitative analysis. SPECT, on the other hand, is much cheaper than PET and does not require the use of an on-site cyclotron since SPECT tracers usually have a much longer half-life than PET tracers, as demonstrated by the approximately 6 h for technetium-99 and or 66 hours for molybdenum-99 against the nearly 2 h for fluorine-18.

2.4 Image Reconstruction

Raw nuclear imaging data only defines the location of an annihilation event on a line, called line integral data. The acquired projection data are mathematically represented by the Radon transform of the raw line integral data. The Radon transform $Rf(\theta, s)$ is a line integral that captures the attenuation or emission of radiation along lines at different angles (θ) and distances (s) from the rotation axis, written as

$$Rf(\theta, s) = \int_{-\infty}^{+\infty} \int_{-\infty}^{+\infty} f(x, y)\delta(x\cos\theta + y\sin\theta - s)dxdy$$

where the integrals are taken over a line of response. The goal of image reconstruction is to use this collected data to create a series of cross-sectional images that represent the object function $f(x, y)$ - the distribution of activity within the subject [1].

A sinogram $(S(\theta, s))$ is a polar plot that encapsulates the entire set of projection data. The sinogram is basically the Radon transform of the object function, modeled as a line integral of count data. The Fourier transform of a line integral at a specific angle in the sinogram's domain is the same as a line through the object in the Fourier domain (Fourier slice theorem). A modified sinogram may be obtained by filtering the transformed data in the Fourier domain, like

$$\hat{S}_f(u, v) = F[S(\theta, s)] \cdot H(u, v)$$

where $H(u, v)$ is a filter and $F[\cdot]$ is the Fourier transform. The filtered sinogram may then undergo inverse Radon transform and the object function is recovered as an image.

The method previously described is known as filtered back-projection and the additional step of filtering is typically performed to enhance high-frequency content of the sinogram to remove radial blurring of simple backprojection. Backprojection in any variety requires that the projection data be complete, which is very rarely the case in modern diagnostic imaging to save on imaging time, and so iterative methods of reconstruction are used [1, 36].

Unlike direct reconstruction methods, iterative techniques refine the image through a series of computational iterations, optimizing the image estimate to match the acquired projection data more closely. The reconstruction process starts with an initial estimate of the image $x^{(0)}$. This could be a simple image or a previous reconstruction. A forward model simulates the imaging process and predicts the expected projection data p_{model} based on the current image estimate $x^{(k)}$

$$p_{model}^{(k)} = A\left(x^{(k)}\right)$$

where $A(\cdot)$ is the system's forward operator and k is the iteration counter. Next, the error is calculated between the acquired projection data and the modeled data

$$r^{(k)} = p_{data} - p_{model}^{(k)}$$

and the model is updated based on this residual difference such that

$$x^{(k+1)} = x^{(k)} + \lambda A^*\left(r^{(k)}\right)$$

where $A^*(\cdot)$ is the system's backprojection operator and λ is a parameter that controls the size of the iterative update. This process is repeated until the difference between updates falls below a threshold [1, 36].

While more specific details of the projection operators and methods of iterative update are beyond the purpose of this text, iterative methods have several notable advantages that come at the cost of computational complexity. Iterative methods are far less sensitive to imaging artifacts, provide better quantitative accuracy for kinetic modeling, and lead to improved image quality through noise suppression [35].

2.5 Overview of Tracers Relevant to Neuroimaging

Tracers are chemically modified radioisotopes designed to bind to specific biological targets. There exists a great variety of tracers which are already approved for use in a variety of applications, or else currently in development. We give a very brief discussion of the tracers specifically useful in nuclear neuroimaging; however, we emphasize that there is a great range of tracers in development (mostly in PET imaging) that measure aspects of brain physiology from synaptic density to neuroinflammation to regional receptor availability, and we narrow our discussion here to those found in clinical practice (or at least approved by the Food and Drug Administration of the USA) rather than research applications.

2.5.1 Perfusion Imaging

Perfusion imaging is a fundamental aspect of nuclear medicine. It revolves around the visualization and evaluation of blood flow dynamics in tissues and organs. SPECT measures blood flow with technetium-99, whereas in PET the assessment of regional blood flow is typically done with oxygen-15 or nitrogen-13 [38–40]. These studies allow for quantitative assessment of regional or organ-specific blood flow (as in brain perfusion studies or myocardial studies) and provide measurements of parameters like regional blood flow, blood volume, and mean transit time [33, 38].

2.5.2 Fluorodeoxyglucose

FDG (2-[^{18}F]-fluoro-2-deoxy-D-glucose) was originally invented for imaging brain metabolism but has since become a staple in oncological imaging [41, 42]. FDG is a glucose analog that may enter neuron cell bodies via glucose transporters but cannot undergo glycolysis [42]. It therefore accumulates in the brain at the rate at which glucose is metabolized; therefore, uptake of FDG is widely recognized as a reliable proxy of synaptic activity and glucose metabolism in general. Changes in local neuronal and synaptic activity have been shown to precede anatomical-structural changes in DSDs [43, 44]. This unique property makes FDG an excellent tracer for the differential diagnosis of DSDs [9, 11, 45–47].

2.5.3 Amyloid and Tau Imaging

Amyloid and tau are two proteins naturally found within the human brain; however, the pathological accumulation and aggregation of these proteins form the biological definition of Alzheimer's disease (AD) under the amyloid cascade hypothesis [48]. In this model, soluble amyloid protein precursors form clumps which dissolve out of the cerebrospinal fluid and aggregate on the surface of the brain. These plaques influence the downstream hyperphosphorylation of cytoskeletal tau proteins in neurons which leads to neuronal injury and cell death. This loss of brain tissue volume is a primary cause of the cognitive decline characteristic of AD.

Tracers which bind to amyloid are used in the early detection and differential diagnosis of AD and other neurodegenerative dementia spectrum disorders [42, 49, 50]. Three different tracers for amyloid aggregates have already been approved for clinical use—florbetaben, florbetapir, and flutemetamol. Amyloid-PET is also a necessary tool for the longitudinal monitoring of therapeutic response to anti-amyloid drugs for the treatment of AD. Amyloid-PET is now entering clinical mainstream and will become increasing common as more anti-AD drugs become available.

Tau tracers, however, exist between clinical reality and research method. One tau tracer has been approved for clinical use [42, 50, 51]. Its use is not nearly as common as those for amyloid, despite extremely promising results in comparison to amyloid or volumetric MRI for monitoring AD progression [52]. Tau-PET may become more common in the future; however, especially if second-generation tau tracers can resolve existing issues with off-target binding and a lack of isoform specificity [51].

3 Topic - Data Preprocessing Operations for Studies in Molecular Neuroimaging

Over the past decade, machine learning (ML) models have widely demonstrated successful application in analyzing intricate patterns within molecular neuroimaging data from various perspectives, including classification [53], regression [54], and severity analysis [13, 55]. A comprehensive ML framework typically consists of five main components: preprocessing of molecular neuroimaging data, feature extraction, feature selection/reduction, the machine learning model itself, and model assessments. It is evident that the effectiveness of an ML framework is closely tied to the specifics of preprocessing, feature extraction, feature selection/reduction techniques, and the choice of machine learning algorithms. In this section, we provide a brief overview of each of preprocessing, feature extraction, and feature selection/reduction techniques steps, while the next section focuses on ML models.

The nature of the preprocessing procedure strongly depends on the specific aim of the following processing algorithm and the type of image. In SPECT, gamma camera detectors capture emitted photons from radiotracers, allowing us to create 3D images that reveal molecular distributions. However, raw SPECT data require meticulous processing to enhance image quality and extract meaningful information [56]. During SPECT imaging, gamma cameras acquire projection data from various angles around the patient. These raw projection images represent the spatial distribution of emitted gamma rays. The aim of data reconstruction is to convert these projections into a 3D image that precisely represents the tracer distribution beneath. The most frequently used reconstruction technique is filtered back-projection, which entails applying a filter to the original projection data to reduce noise and artifacts. Other methods of reconstruction include Iterative Reconstruction algorithms, such as maximum likelihood expectation maximization (MLEM) or ordered subset expectation maximization (OSEM), which continuously refine the reconstructed image. These algorithms consider system response, attenuation, scatter, and resolution effects. While iterative reconstruction enhances image quality, it necessitates more computational resources. After reconstruction, an Image Enhancement is applied to improve the quality of the reconstructed SPECT image. The main techniques for enhancing SPECT images include attenuation correction, scatter correction, and resolution recovery. The subsequent stage includes measuring tracer uptake, which involves assessing the quantity of radiotracer uptake. This is commonly achieved through metrics like standardized uptake value ratios (SUVR) or percent injected dose per gram (%ID/g). These values provide valuable insights into functional characteristics of the brain, such as blood flow activity in the brain [57]. SPECT imaging plays a significant role in different scenarios, including psychiatric disorders, substance abuse, and neurodegenerative diseases [58, 59].

When it comes to PET images, the initial steps involve image acquisition and reconstruction, which are essential for processing PET data. During these stages, PET scanners detect emitted positrons from the radioactive tracer and create detailed 3D images that show the brain's metabolic activity. Next, corrections for distortions and artifacts are applied to improve the accuracy and quality of the PET data. Then, the PET data underwent a process of standardization and normalization techniques. These steps account for inter-individual variability in anatomy and radiotracer distribution. By standardizing the data and normalizing it to a common template, the variability between individuals is accounted for. Quantification and noise reduction are essential components of PET data preprocessing, requiring careful techniques to minimize noise and enhance the precision of molecular distribution in the brain. Finally, the PET images were converted to SUVR [60].

PET imaging can use a variety of radiotracers that are customized for specific diseases or conditions [61]. Fluorodeoxyglucose (FDG) is frequently utilized as a radiotracer in PET imaging to assess glucose metabolism in the brain. This specific type of PET imaging capitalizes on the brain's high demand for glucose, enabling the distinction of different brain pathologies and the early identification of Alzheimer's disease [62]. PET imaging can also be performed using specific tracers, like 18F-fallypride, which is becoming more valuable in clinical research and drug development. These tracers need to meet certain requirements, such as being able to enter the brain effectively and reversibly without interference from problematic radiometabolites, allowing for precise measurement of the target protein density in the brain [63].

Once the preprocessing is complete, feature extraction is used to provide functional brain features as input for a machine learning model. This can be achieved through either a whole-brain strategy, which considers all brain voxels, or a region-wise strategy that involves segmenting the brain image into different regions and calculating the average functional activity of each region as a brain feature. One major limitation of utilizing a whole-brain approach for feature extraction is the issue of dimensionality. Consequently, it may be necessary to implement feature selection or reduction techniques to combat this challenge.

In neuroimaging-based ML studies, principal component analysis (PCA) is a frequently employed data reduction method [64, 65]. It is worth noting that PCA is considered an unsupervised feature reduction technique, meaning that it reduces the input space without enhancing prediction accuracy.

4   Topic - Machine Learning Methods

The objective of a machine-learning model is to uncover a correlation between a dependent variable and its independent variables in the training dataset, to predict outcomes on unseen test datasets. There exist four broad categories of learning paradigm: supervised, unsupervised, semi-supervised, and reinforcement based.

Supervised learning is a cornerstone of machine learning where models are trained on a labeled dataset. In this paradigm, the algorithm learns to map input features to corresponding target labels. The objective is to generalize this learned mapping to accurately predict labels for new, unseen data. Common applications include classification and regression tasks, where the algorithm categorizes data or predicts numerical values based on labeled examples. Supervised learning finds extensive application in healthcare, contributing to tasks such as disease diagnosis and predictive analytics. For instance, a supervised learning model can be trained on labeled medical images to identify patterns indicative of specific conditions or used for predicting patient outcomes based on historical data [66].

Unsupervised learning deals with unlabeled data, aiming to uncover inherent patterns and structures without explicit guidance. Clustering and dimensionality reduction are common tasks in this paradigm. The algorithm explores the data's intrinsic characteristics, revealing hidden relationships and groupings in large and high-dimensional datasets. In healthcare, unsupervised learning can be applied to cluster patients based on similar health profiles or detect anomalies in large datasets, contributing to personalized medicine and improving outlier detection [66].

Hybrid approaches combine elements of supervised and unsupervised learning. Semi-supervised learning, for instance, utilizes a small amount of labeled data alongside a larger pool of unlabeled data. This paradigm seeks to strike a balance, harnessing the advantages of both labeled and unlabeled information. Unsupervised pretraining may also be used to aid in initializing a model's parameters before fine-tuning with supervised training. This practice can also address data scarcity in small datasets [66–68]. Reinforcement learning involves agents making decisions in an environment to maximize cumulative rewards. This paradigm is particularly relevant in healthcare decision-making, where models learn optimal strategies by interacting with the environment [69]. In this chapter, we focus on the supervised machine-learning algorithms, specifically classification and regression models, used in molecular-based ML studies.

4.1 Classification Models

In molecular brain imaging research, the primary classification techniques widely used include support vector machine (SVM), random forest (RF), neural networks (NN), and deep learning methods. SVM is a powerful classifier based on statistical learning principles, known for its effectiveness in neuroimaging-based classification tasks. It is particularly able to solve complex problems, especially when dealing with high-dimensional data and studies with limited samples. A comprehensive exploration of deep learning concepts, structures, and operations is provided in the subsequent section (Subheading 5).

4.2 Regression Models

When it comes to regression models in neuroimaging studies, support vector regression (SVR) stands out as the most widely employed regression model. Like linear regression, SVR strives to identify an optimal hyperplane that minimizes deviations from the training data. Gaussian process regression (GPR) is another well-known regression model that is a nonparametric Bayesian method used for regression tasks. It operates based on a probability distribution that covers potential values [65, 70]. Additional details on alternative classification and regression models utilized in the field of neuroimaging can be referenced elsewhere [71, 72].

5 Topic - Deep Learning Methods

5.1 Artificial Neurons: An Analogy to Nature

Artificial neurons are the elemental unit of deep learning, inspired by the complexity observed in biological neural systems. These digital entities, mirroring their biological counterparts, assume a pivotal role in shaping the functionality and learning dynamics of neural networks through their plasticity and ability to model nonlinear operations [67].

Functioning as information processors at their core, artificial neurons emulate the electrochemical signaling inherent in biological neural networks. Analogous to dendrites collecting signals from neighboring cells in biological neurons, artificial neurons receive input from diverse sources. These inputs undergo a weighting process, reflecting the significance assigned to each signal in the decision-making process. The weighted inputs undergo summation and biasing

$$z = \sum_{i=1}^{n} w_i \cdot x_i + b$$

before passing through an activation function which determines the firing or activation status of the neuron based on the amalgamated input [66, 67]. These weights and biases are called learnable parameters and undergo adjustments during the network's training phase. Through iterative exposure to training data, the network fine-tunes these parameters, adapting its responses to specific patterns and features present in the input data.

The activation function, resembling the threshold in biological neurons, acts as a digital decision-maker, dictating the transmission or inhibition of signals. This binary response forms the foundation of computational decision-making within the neural network. The output from this process serves as the input for subsequent layers of neurons, initiating a cascading effect that characterizes the layered architecture inherent in artificial neural networks, also called ANNs [73].

Various types of activation functions exist, each with unique characteristics and applications. The sigmoid activation function compresses input values to a range between 0 and 1, rendering it suitable for binary classification tasks

$$a(z) = \frac{1}{1 + e^{-z}}.$$

The hyperbolic tangent (tanh) function extends this range from -1 to 1, providing a broader spectrum for representation

$$a(z) = \frac{e^z - e^{-z}}{e^z + e^{-z}}.$$

Rectified linear unit (ReLU) has gained popularity for its computational simplicity and practical effectiveness, allowing positive values to pass through while zeroing out negative values

$$a(z) = \max(0, z).$$

The choice of activation function influences the network's learning capacity, convergence speed, and ability to handle different data types.

5.2 Nodes, Layers, and Networks

Neural networks are characterized by their layered composition with information flowing through interconnected nodes. A layer is a collection of nodes (neurons), which collectively process and transform the input data. The network is typically organized into three main types of layers: input layers, hidden layers, and output layers.

Input layers receive the initial input data and serve as the entry point for information into the network. Each node in the input layer represents a feature or attribute of the input data. Hidden layers lay between input and output and neural networks have a single or several thousand hidden layers depending on their size and complexity. Nodes in these layers perform complex computations of the input data to capture the patterns and representations within. The depth and width of hidden layers contribute to the network's capacity to learn hierarchical features. Output layers are the final layer within a network and produce the network's output, usually a prediction or classification. The number of nodes in the output layer depends on the nature of the task, such as binary or multiclass classification, regression, or other specific objectives [66, 67].

The output of a layer can be expressed as

$$a^{(l)} = f\left(W^{(l)} \cdot a^{(l-1)} + b^{(l)}\right)$$

where $a^{(l)}$ is the output of layer (l), $f(\cdot)$ is the activation function, $W^{(l)}$ is the matrix of weights for a layer, $a^{(l-1)}$ is the input from the previous layer, and $b^{(l)}$ is the bias term for a layer. Where the function of weights is multiplicative, the bias terms account for constant offsets by allowing for correct modeling of non-centered data. The interconnected nodes within layers communicate through weighted connections, with each connection representing a parameter that the network learns during training.

5.3 Forward and Backward Passes

The feedforward pass is the initial phase where input data traverses through the network, layer by layer, culminating in the generation of an output at the final layers of the neural network. The iterative nature of the feedforward pass ensures that each layer refines the representation of the input data, capturing increasingly complex features.

At the point of output, a comparison with prediction and a measure of said prediction must be made. This is done by use of a loss function [74]. The loss function serves as the compass guiding the network toward optimal parameter values. It quantifies the disparity between the predicted output and the actual target values and provides a basis for assessing the performance of a neural network. The general expression for a loss function is written as

$$\mathcal{L} = \mathrm{Loss}(\hat{y}, y)$$

where $\hat{y}$ and y are the true targets and predicted outputs, respectively.

Common choices of loss function are the mean squared error (MSE) for regression tasks

$$\mathcal{L}_{MSE} = \frac{1}{N} \sum_{i=1}^{N} (y_i - \hat{y}_i)^2$$

and cross-entropy for classification tasks

$$\mathcal{L}_{CE} = -\frac{1}{N} \sum_{i=1}^{N} \left[\hat{y}_i \log(y_i) + (1 - \hat{y}_i) \log(1 - y_i)\right]$$

where N is the number of samples and the subscript i is the sample index.

Following the feedforward pass and the computation of the loss, backpropagation takes center stage in the iterative process of refining the network's performance. This crucial step involves the calculation of gradients and the adjustment of weights and biases based on the error incurred during the feedforward pass. The backpropagation algorithm utilizes the chain rule of calculus to compute the gradients of the loss function with respect to the weights and biases [66, 67]. These gradients guide the network in updating its parameters to minimize the error. The process involves the backward pass through the layers, propagating the error information from the output layer back through the network to the input layer. The update of parameters during backpropagation can be expressed as

$$W_{t+1}^{(l)} = W_t^{(l)} - \alpha \frac{\partial L}{\partial W_t^{(l)}}$$

$$b_{t+1}^{(l)} = b_t^{(l)} - \alpha \frac{\partial L}{\partial b_t^{(l)}}$$

where α is the learning rate and the terms $\frac{\partial L}{\partial W_t^{(l)}}$ and $\frac{\partial L}{\partial b_t^{(l)}}$ denote the partial derivatives of the loss function with respect to the matrices of weights and biases, respectively. It is common to suppress biases and simply include them as entries in the weight matrices; however, they are explicitly stated here for conceptual clarity.

This method of backpropagation is known as gradient descent. Computing the partial derivatives is equivalent to computing the gradient of the loss function

$$W_{t+1}^{(l)} = W_t^{(l)} - \alpha \nabla L \left(W_t^{(l)} \right)$$

where we assume the gradient is computed with respect to the parameters of the network. An important property of the gradient is that it points along a direction of maximum change in the function to which it is applied. The iterative cycling of feedforward and backpropagation refines the network's parameters such that these may eventually minimize the overall loss.

This process is, however, also affected by the choice of non-learnable parameters called hyperparameters. These are externally provided and govern the behavior of the network. These include the learning rate, batch size, degree of regularization, and number of nodes and layers. The selection of hyperparameters is a largely heuristic process that occurs during model development and ultimately influences the convergence speed, generalization capacity, and overall performance of the neural network.

5.4 Convolutional Neural Networks

ANNs have proven to be powerful tools for various tasks, yet when confronted with grid-like data such as images, they encounter challenges in effectively capturing spatial relationships. The densely connected layers of ANNs lack the ability to exploit the inherent grid structure of images. Each neuron in a fully connected layer considers the entire input, leading to an overwhelming number of parameters and computational inefficiencies. Convolutional neural networks (CNNs) leverage convolutional filters in a specialized architecture tailored for grid-based data like images. Convolutional filters allowed CNNs to learn hierarchical representations from low level (simple edges) to high level (complex nonlinear features) by systematically scanning local regions to extract features with far greater computational efficiency than using a conventional ANN. The hierarchical abstraction facilitated by convolutional filters aligns with the hierarchical nature of visual information in mammalian vision, allowing CNNs to excel in tasks involving grid-like data [66, 67, 75].

Convolution involves applying a filter (AKA kernel) onto the input image and computing the dot product at each spatial location. This process results in a feature map, highlighting patterns and structures present in the input. Filters act as the eyes of convolutional operations. These small, learnable matrices are responsible for detecting specific features within the input [75].

The convolution between an image I and a filter K can be expressed as

$$S(i,j) = (I * K)(i,j) = \sum_{m}^{M} \sum_{n}^{N} I(m,n) \cdot K(i-m, j-n)$$

where $S(i,j)$ represents the value at position (i,j) of the resulting feature map, $I(m,n)$ is the pixel intensity of the input image at position (m,n), and $K(i-m, j-n)$ is the filter's weight at relative position $(i-m, j-n)$ where the operation is performed over the entire image (see Fig. 4 for an illustration). Filters weights are also shared such that one filter is applied over an image instead of each local region having its own filter. Parameter sharing significantly reduces the number of learnable parameters compared to fully connected layers in ANNs.

Like ANNs, CNNs also make use of forward and backward passes, and in conjunction with computing gradients of the loss function, will iteratively adjust the filter weights by backpropagation to improve performance.

To enhance efficiency and reduce computational load, CNNs incorporate pooling layers. Pooling layers facilitate downsampling, reducing the spatial dimensions of the input while preserving essential features. Max pooling, for instance, selects the maximum value from a group of neighboring pixels, emphasizing the most salient information and making the network less sensitive to small changes

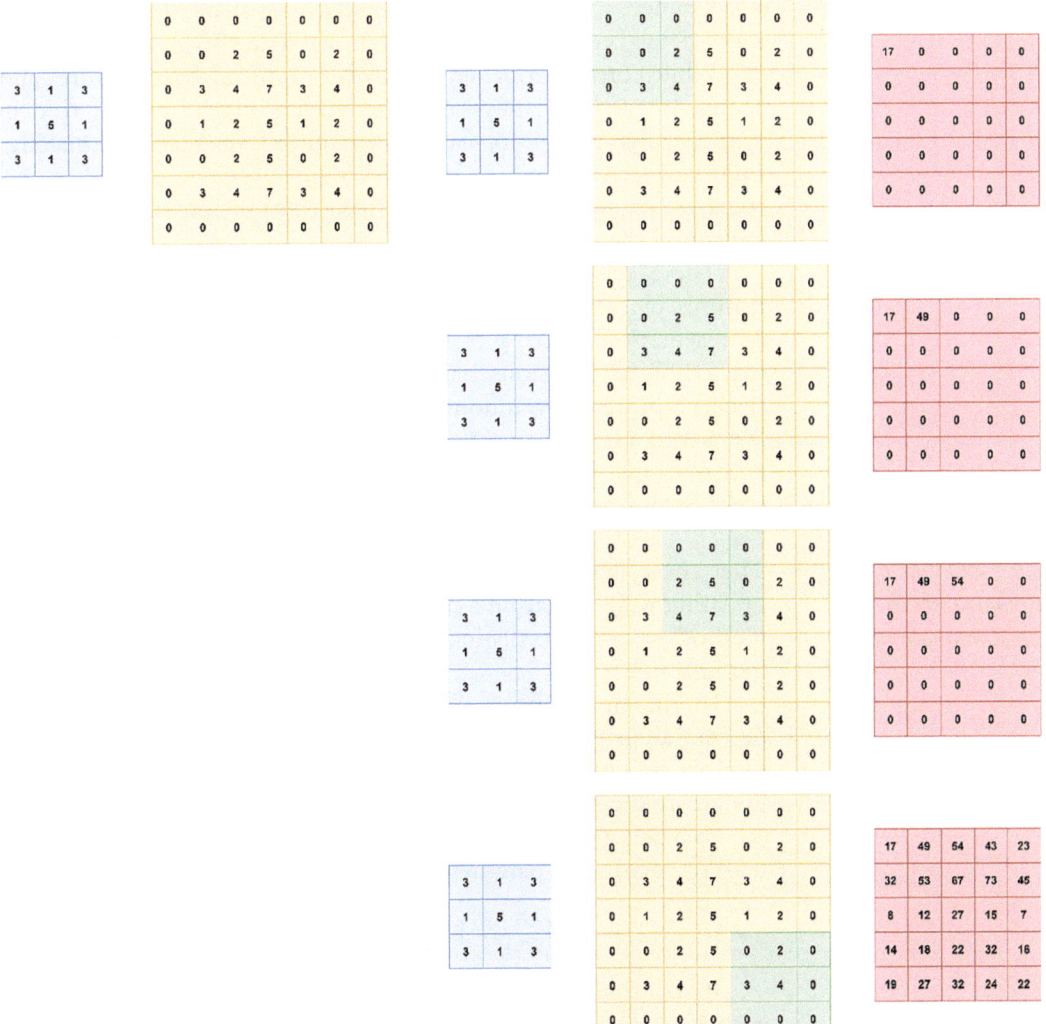

Fig. 4 Convolutional filter in two dimensions. The kernel/filter in blue passes over the image data in yellow one place at a time over the entire image. Note that the image has been zero padded to produce a 7 × 7 matrix. This is commonly performed to prevent aggressive reduction in dimensionality by convolutional filtering. As the filter passes over the image, the filter weights multiply the values within the image data, and the result is summed as an output at the center of the kernel to create an output image in red

in data. This downsampling not only expedites computation but also enhances the network's translational invariance, allowing it to recognize features regardless of their precise spatial location within an image [75].

Both convolutional and pooling operations may also skip over a certain number of pixels when performing their operations. This is known as stride. Strides dictate the step size of the convolutional kernel as it traverses the input. A larger stride leads to greater downsampling and computational efficiency but may sacrifice fine-grained information. Conversely, a smaller stride preserves more detail but increases computational load [75].

While convolutional layers excel at capturing local features and spatial hierarchies, fully connected layers play a crucial role in synthesizing this information for high-level decision-making. Fully connected layers, commonly found at the end of CNN architectures, connect every neuron from the previous layer to every neuron in the subsequent layer.

These layers act as global aggregators, assimilating the localized features extracted by convolutional layers into a comprehensive representation. This global perspective allows the network to make holistic decisions based on the amalgamation of intricate details captured throughout the image. Integrating fully connected layers into a CNN transforms the hierarchical features extracted by convolutional layers to simple numbers such that their nonlinear combinations are useful for performing regression or classification.

5.5 Methods of Optimization in Backpropagation

There exist many different optimization algorithms for iteratively updating weights within a neural network; however, the most used methods create two distinct categories: nonadaptive and adaptive method.

In the nonadaptive category, we have the vanilla gradient descent algorithm and its extensions. These differ primarily in the number of samples used to compute loss gradients, the stability of update gradients, and computational stability.

In batch gradient descent (BGD), loss gradients are computed with respect to the parameters for the entire dataset, given as

$$W_{t+1}^{(l)} = W_t^{(l)} - \alpha \nabla \mathcal{L}_{\text{batch}}\left(W_t^{(l)}\right).$$

By considering the global information from all data points in each iteration, BGD aims to converge to the optimal set of parameters that minimize the overall loss. This approach provides a precise direction for optimization but has tremendous computational requirements, especially for large datasets. BGD may be used in scenarios where the entire dataset can comfortably fit into memory and computational efficiency is not a primary concern [76].

Stochastic gradient descent (SGD) takes a departure from BGD by randomly selecting a single data point for gradient computation in each iteration, written as

$$W_{t+1}^{(l)} = W_t^{(l)} - \alpha \nabla \mathcal{L}_i\left(W_t^{(l)}\right)$$

where the subscript i is the randomly chosen datapoint. This randomness introduces a level of noise that can help escape nonoptimal local minima and navigate the loss function landscape more flexibly. The effect of the randomness is that the optimization minimizes the expected value of the losses over the dataset. SGD is particularly useful when dealing with vast datasets, as it significantly reduces the computational burden associated with processing the entire dataset, as would be the case in BGD; however, the stochastic nature of the updates can lead to erratic convergence and oscillations in the loss function [76].

Mini-batch gradient descent (MGD) strikes a balance between BGD and SGD by utilizing a subset, or mini-batch, of the dataset for each iteration, given as

$$W_{t+1}^{(l)} = W_t^{(l)} - \alpha \nabla \mathcal{L}_{\text{mini-batch}}\left(W_t^{(l)}\right).$$

This method combines the advantages of leveraging global information while enjoying the computational efficiency gained from processing smaller batches. The mini-batch size becomes a crucial hyperparameter that influences the optimization process. MGD is widely adopted in practice, providing a practical compromise that suits various dataset sizes and computational constraints, but is sensitive to the choice of hyperparameters [76].

Nesterov accelerated gradient descent (NAG) is an extension of SGD that introduces a look-ahead mechanism to address the challenge of momentum in gradient descent, with the update rule

$$W_{t+1}^{(l)} = W_t^{(l)} - \alpha \nabla \mathcal{L}\left(W_t^{(l)} + \beta\left(W_t^{(l)} - W_{t-1}^{(l)}\right)\right)$$

where β represents the momentum parameter. NAG computes the gradient at the anticipated future position of parameters [77]. This anticipation allows NAG to reduce oscillations and converge faster compared to standard SGD. NAG effectively combines the exploration benefits of SGD with a form of momentum, allowing it to take advantage of smooth loss surfaces and navigate complex ones with sharp turns [76].

In the adaptive category, we have more sophisticated methods that dynamically adjust learning rates during the training process, which typically results in more efficient convergence.

AdaGrad, an adaptive gradient algorithm, adjusts the learning rates for each parameter based on historical gradients. The update rule for AdaGrad is given by

$$W_{t+1}^{(l)} = W_t^{(l)} - \frac{\alpha}{\sqrt{G_t^{(l)} + \epsilon}} \nabla \mathcal{L}\left(W_t^{(l)}\right)$$

where $G_t^{(l)}$ is the accumulated squared gradient and ϵ is a small constant for numerical stability. AdaGrad effectively adapts the learning rates, providing larger updates for infrequent parameters and smaller updates for frequently occurring ones; however, it has limitations in scenarios where the accumulated gradients become large, leading to overly conservative updates [78].

Root mean square propagation (RMSProp) addresses Ada-Grad's issue of aggressive, monotonically decreasing learning rates with the refined update rule

$$W_{t+1}^{(l)} = W_t^{(l)} - \frac{\alpha}{\sqrt{E\left[G^{(l)}\right]_t + \epsilon}} \nabla_L\left(W_t^{(l)}\right)$$

where $E[G^{(l)}]_t$ is the expected value of the average of squared gradients. This function has the notable property that it decays exponentially and so RMSProp mitigates the problem of rapidly diminishing learning rates by considering a moving average. This allows the algorithm to adapt to varying gradients during different stages of optimization, but like AdaGrad, RMSProp may still suffer from vanishing learning rates in certain situations [76].

Adam (adaptive moment estimation) combines the benefits of both momentum and RMSProp, introducing adaptive moment estimates for each parameter. The update rule for Adam is given by

$$W_{t+1}^{(l)} = W_t^{(l)} - \frac{\alpha}{\sqrt{V_t^{(l)} + \epsilon}} M_t^{(l)}$$

where $V_t^{(l)}$ is the exponentially decaying average of squared gradients and $M_t^{(l)}$ is the exponentially decaying average of gradients. Adam excels in scenarios with nonstationary objectives and noisy gradients, combining the benefits of adaptive learning rates and momentum. However, it comes with the challenge of sensitivity to hyperparameters and potential overfitting in certain cases due to issues with bias correction [79].

The difference between traditional gradient descent-based methods and adaptive methods lies in their approach to adjusting learning rates during optimization. While traditional methods maintain a fixed learning rate for all parameters, adaptive methods dynamically adjust learning rates based on historical gradient information. The former offers stability and straightforwardness but the latter excel in handling varying gradients, offering faster convergence and improved performance on nonstationary objectives. Adaptive methods also introduce additional hyperparameters.

5.6   Overfitting and Validation Strategies

5.6.1   Understanding Goodness of Fit

Overfitting and underfitting are critical challenges where the goal is to create models that generalize well to unseen data Overfitting occurs when a model not only learns the underlying patterns in the training data but also captures noise and peculiarities specific to that dataset. As a result, the model performs exceptionally well on the training set but fails to generalize with new unseen data. On the other hand, underfitting happens when the model is too simplistic to capture the underlying patterns, resulting in poor performance on both the training and test sets. Achieving a well-fit model is the ultimate objective and denotes a balance where the model comprehensively understands the underlying structures of the data, enabling it to perform consistently well on both training and testing datasets. Striking this balance is crucial for building models that generalize effectively and produce meaningful predictions.

5.6.2   Regularization Techniques

Regularization methods are a group of techniques used to mitigate overfitting by imposing constraints on the model's parameters, preventing them from adapting too closely to the idiosyncrasies of the training data [80, 81].

L_1 encourages sparsity in the model by adding a penalty proportional to the absolute values of the weights

$$R_{L_1} = \lambda \sum_{i=1}^{n} |W_i|$$

where λ is the regularization constant and controls the strength of the regularization term [67]. This penalty promotes a sparse selection of features, effectively ignoring irrelevant ones, preventing the model from becoming too intricate or bespoke to the training data.

L_2 regularization (AKA weight decay) penalizes the sum of squared weights

$$R_{L_2} = \lambda \sum_{i=1}^{n} W_i^2.$$

It discourages overly large weights, preventing the model from becoming overly sensitive to small variations in the training data [67]. Both L_1 and L_2 regularizations contribute to a more stable and well-behaved optimization landscape by adding penalties into the loss function which discouraging the model from fitting noise and promotes the learning of essential patterns.

Dropout is a regularization technique unique to deep learning models that introduces stochasticity during training by randomly deactivating a fraction of neurons within a network [81]. This prevents the network from relying too heavily on specific neurons, forcing it to distribute learning across a broader range of features. The utility of dropout lies in its ability to act as an implicit ensemble method, producing multiple slightly different models. During inference, the predictions from these sub-networks are averaged, resulting in a more robust and generalized output. Dropout, therefore, not only reduces overfitting by preventing co-adaptation of neurons but also promotes model robustness and generalization.

5.6.3   Validation Strategies

Validation and testing sets play crucial roles in assessing the performance of machine models and ensuring their generalization to unseen data [82]. The training set is used to train the model, allowing it to learn patterns from labeled data, but to evaluate how well the model performs on data it has never seen before: validation and test data.

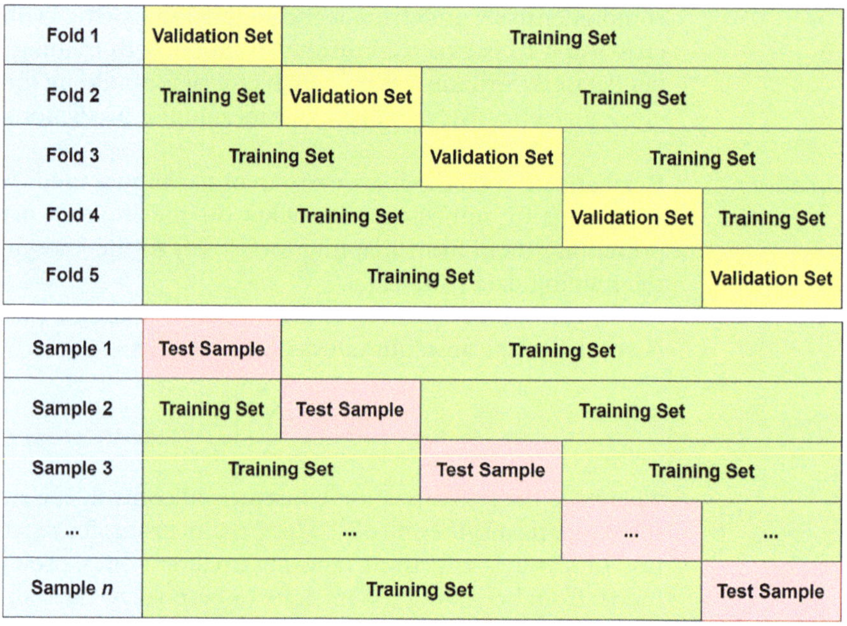

Fold 1	Validation Set	Training Set			
Fold 2	Training Set	Validation Set	Training Set		
Fold 3	Training Set		Validation Set	Training Set	
Fold 4	Training Set			Validation Set	Training Set
Fold 5	Training Set				Validation Set

Sample 1	Test Sample	Training Set			
Sample 2	Training Set	Test Sample	Training Set		
Sample 3	Training Set		Test Sample	Training Set	
...	...		...	...	
Sample n	Training Set				Test Sample

Fig. 5 Graphical comparison of K-fold cross-validation (KCV) and leave-one-out cross-validation (LOOCV) (above). In this example, we see a cross-validation where the data have been split into fivefolds. From here, the average error or prediction may be computed to estimate performance over the entire data set. The same principle is applied here except that the data are divided into folds equal to the number of samples, making the validation set equivalent to a single sample

The validation data are used during the training process as a separate dataset to evaluate the model's performance and provides a measure of overfitting to the training data. A model that performs well on training data but poorly on validation data is are likely overfit. Validation data also provides an avenue to tuning hyperparameters during model development.

The true test of its generalization capabilities comes with the testing set. The testing set serves as the final arbiter, providing an unbiased evaluation of the model's performance on completely new and unseen data. By withholding this data during the training and validation phases, the testing set ensures an objective assessment, reflecting how well the model is likely to perform in real-world scenarios.

Several means of partitioning a dataset into training, validation, and testing sets exist. K-fold cross-validation (KCV) and leave-one-out cross-validation (LOOCV) are the most used paradigms. *See* Fig. 5 for a comparison between the two.

KCV involves dividing the dataset into K equally sized folds. The model is trained K times, each time using $K - 1$ folds for training and the remaining fold for validation. This process is repeated K times, with each fold serving as the validation set exactly once. The final performance metric is the average of the

K evaluations. This provides a more robust estimate of a model's performance by leveraging multiple train-test splits, reducing the impact of variability in a single split and offering a more representative assessment of generalization [82].

LOOCV takes this idea to the extreme by using each individual data point as a separate validation set. For a dataset with N samples, LOOCV involves N iterations, training the model N times, leaving out one sample as the validation set in each iteration. LOOCV provides an unbiased estimate of model performance as it uses almost all available data for training in each iteration, but it can be computationally expensive for large datasets [82].

The utility of KCV and LOOCV lies in their ability to provide a more reliable estimate of a model's generalization performance. By systematically cycling through different subsets of the data for training and validation, these techniques offer a comprehensive evaluation that captures the model's ability to perform well on diverse subsets of the dataset [82]. Other variants such as the train-validation-test split are also reasonable choices, and refinements such as nested cross-validation also exist.

5.7 Important Neural Network Architectures

We now turn our attention to a survey of important architectures in deep learning networks. Each of these introduces a new paradigm to the field to address specific challenges and harness distinct capabilities.

5.7.1 Residual and Dense Connectivity

Residual neural networks (ResNets) emerged to tackle the complexities of training exceptionally deep networks. With increasing depth, the vanishing gradient problem hampers effective learning. ResNets leverage residual connections, or skip connections, to enable smoother gradient flow by allowing information to bypass certain layers, which promotes convergence in deep neural networks. The identity mapping alongside the learned mapping in each block facilitates training by mitigating the vanishing gradient problem. Mathematically, a residual block is expressed as

$$H_{t+1} = H_t + F(H_t, W_t)$$

where H_t is the input, W_t are learnable parameters, and $F(\cdot)$ is the residual function. This innovative architecture empowers the training of extremely deep networks, capturing complex hierarchical features. The motivation behind the architecture of these networks to learn the difference (residuals) between the input and output of a residual block [83].

Densely connected neural networks (DenseNets) are architectures where the idea of skip connections is taken to the logical end by introducing dense connectivity patterns where each layer receives direct inputs from all preceding layers within a dense block; however, instead of focusing on learning residuals, the concern is with feature reuse and information flow. To these ends, DenseNets use depth-wise concatenation of feature maps

$$H_{t+1} = [H_0, H_1, H_2, ..., H_n] + F(H_t, W_t)$$

Features between dense blocks within this architecture may be randomly bottlenecked or compressed between dense blocks to decrease computational demand and promote feature independence [84].

5.7.2 Feature Compression and Unsupervised Learning

Autoencoders are a neural architecture designed to learn efficient representations of input data. This ability to capture essential features and patterns makes autoencoders valuable for tasks like dimensionality reduction, anomaly detection, and generative modeling. Unlike supervised learning, where models are trained with labeled data, autoencoders fall under the category of unsupervised learning, making them adept at extracting meaningful information from unlabeled datasets [85].

The architecture consists of an encoder and a decoder, with the goal of reconstructing the input data through a bottleneck or latent space. The encoder takes the input data and maps it to a lower-dimensional latent space, effectively compressing the information, and the decoder then reconstructs the input data from this reduced representation. The training objective revolves around minimizing the reconstruction error, compelling the autoencoder to learn a compact and informative representation of the data.

Let us denote the input data as X, the encoder function as E, the decoder function as D, and the reconstructed data as X'. The encoding and decoding processes can be expressed as

$$Z = E(X) = f(W_e X + b_e)$$

where Z represents the latent representation or encoding of the input data, W_e and b_e are the weights and biases of the encoder, respectively, and $f(\cdot)$ is an activation function. The decoder then aims to reconstruct the input from this latent representation

$$X' = D(Z) = f(W_d Z + b_d)$$

and often incorporates transposed convolutional layers, effectively reversing the transformations of the encoder and upsampling the latent space expression. The objective during training is to minimize the difference between the input data and its reconstruction. This is achieved through minimizing a loss function such as the MSE.

Variational autoencoders (VAEs) introduce probabilistic elements into the encoding and decoding process, transforming autoencoders into generative models [86]. The encoder outputs not only the mean μ of the latent distribution but also its variance σ^2. Rather than directly sampling from this distribution, a reparameterization trick is employed, introducing a new variable ϵ sampled from a learned distribution (usually Gaussian). The sampled latent vector z is then obtained by transforming the mean and variance according to

$$z = \mu + \epsilon \cdot \sigma$$

This stochasticity in the encoding process enables VAEs to generate diverse samples during both training and inference. The decoder, too, is imbued with a probabilistic nature, producing data with inherent uncertainty.

A key component of the formulation of a VAE is the addition Kullback-Leibler (KL) divergence term added to the loss function. This additional term regularizes the latent space, encouraging it to follow a specific distribution. The KL divergence between the learned distribution and the target distribution is given by

$$KL[q(z|x), p(z)] = -\frac{1}{2} \sum_{i=1}^{K} \left(1 + \log(\sigma_i^2) - \mu_i^2 - \sigma_i^2\right)$$

where $q(z|x)$ is the learned distribution and $p(z)$ is the target distribution. The use of the KL divergence and the reparameterization trick makes VAEs more expressive, allowing them not only to faithfully reconstruct input data but also to generate novel samples that capture the underlying structure of the data distribution in the latent space.

5.7.3 Generative Models

Generative adversarial networks (GANs) are neural architectures designed for the explicit purpose of generating realistic data samples [87]. In contrast to autoencoders, which focus on learning efficient representations of input data, GANs are centered around the adversarial training of two neural networks. The GAN architecture involves a generator (G) and a discriminator (D). The generator takes random noise as input and transforms it into data samples, attempting to create realistic output. Simultaneously, the discriminator evaluates the generated samples along with real ones, aiming to distinguish between genuine and synthetic data. Similar to autoencoders, the generator network also uses transposed convolutional layers for upsampling from sampled noise. The adversarial objective drives the generator to improve its ability to generate realistic samples, while the discriminator refines its capacity to differentiate between real and generated data.

The generator's objective is to generate data that is indistinguishable from real samples, which is expressible as a minimax game on a value function

$$\min_{G} \max_{D} V(D, G) = \mathbb{E}_{x \sim p_{\text{data}}(x)} [\log(D(x))]$$

$$+ \mathbb{E}_{z \sim p_{\text{noise}}(z)} [\log(1 - D(G(z)))]$$

where $p_{\text{data}}(x)$ and $p_{\text{noise}}(z)$ are the distributions of real data and random noise respectively, $\mathbb{E}$ is the expected value operator, $G(z)$ is a generated sample, and $D(x)$ is the discriminator's output.

The goal of the GAN training is to find a generator that minimizes this value function while simultaneously finding a discriminator that maximizes it. These iteratively improve their performance through adversarial training. Equilibrium is reached when the generator produces realistic samples that the discriminator cannot distinguish from real data, and the discriminator is effectively guessing whether the input is real or generated with a 50% probability.

Conditional GANs (cGANs) extend the capabilities of traditional GANs by introducing conditional information during the training process [88]. In a cGAN, both the generator and discriminator receive additional information, typically in the form of class labels or other auxiliary data, to guide the generation process. This conditioning allows for the targeted generation of samples belonging to specific classes or exhibiting desired characteristics.

The objective function for a cGAN is modified to include conditional information such that

$$\min_{G} \max_{D} V(D,G) = \mathbb{E}_{x \sim p_{\text{data}}(x,y)}[\log D(x|y)]$$

$$+ \mathbb{E}_{z \sim p_{\text{noise}}(z)}[\log(1 - D(G(z|y)))]$$

where $p_{\text{data}}(x,y)$ is instead the joint distribution of real data and conditional labels, $D(x|y)$ is the discriminator's output given real data and conditional labels, and $G(z|y)$ the generator's output based on noise and conditional labels.

5.7.4 Sequential/ Temporal Architectures

Recurrent neural networks (RNNs) are specifically designed to tackle the challenges posed by sequential data (spatial or temporal). Unlike feedforward networks, RNNs introduce a dynamic element by incorporating feedback loops, allowing them to maintain a memory of previous inputs in their internal state [89].

The key feature of RNNs lies in their ability to capture dependencies across time steps. Let us denote the input at time step as X_t, the hidden state as H_t, and the output as Y_t. The recurrent nature is encapsulated in the hidden state, which is updated based on both the current input and the previous hidden state with the rule

$$H_t = f(W_{hh}H_{t-1} + W_{xh}X_t)$$

where W_{hh} represents the weight matrix for hidden-to-hidden connections and W_{xh} denotes the weight matrix for input-to-hidden connections. The activation function $f(\cdot)$ introduces nonlinearity to this relation.

The dynamic unfolding of this recurrent structure over multiple time steps creates a network capable of processing sequential information; however, it must be noted that standard RNNs face challenges like vanishing gradients that limit their ability to capture long-range dependencies.

Long short-term memory networks (LSTMs) emerge as a crucial innovation in addressing fundamental challenges encountered by traditional RNNs, primarily the vanishing gradient problem. LSTMs introduce a sophisticated memory cell structure that enables the network to selectively store or discard information, preventing the vanishing gradient problem [90].

A memory cell contains input, output, and forget gates. The input gate i_t, forget gate f_t, and output gate o_t control the flow of information in, within, and out of a memory cell state C_t with the update rule

$$H_t = o_t \tanh(C_t)$$

where the cell state C_t is written as

$$C_t = f_t \cdot C_{t-1} + i_t \cdot \tilde{C}_t$$

and $\tilde{C}_t$ is the candidate memory cell state

$$\tilde{C}_t = \tanh\left(W_{ig}X_t + W_{hg}H_{t-1} + b_g\right)$$

and i_t, o_t, and f_t are the input, output, and forget gates written as

$$i_t = \sigma\left(W_{ii}X_t + W_{hi}H_{t-1} + b_i\right)$$

$$o_t = \sigma\left(W_{io}X_t + W_{ho}H_{t-1} + b_o\right)$$

$$f_t = \sigma\left(W_{if}X_t + W_{hf}H_{t-1} + b_f\right)$$

respectively. Note that $\sigma(\cdot)$ is the sigmoid activation function specifically.

5.7.5 Segmentation Architectures

The U-Net architecture (named for its distinctive shape) emerged as a solution to the challenges posed by semantic segmentation tasks in medical image analysis, particularly in scenarios where the structures of interest exhibit intricate shapes and patterns.

The primary motivation behind U-Net's creation was the inadequacy of traditional convolutional neural networks (CNNs) for pixel-wise segmentation tasks. Standard CNN architectures suffered from the loss of spatial resolution during the downsampling process, making it challenging to precisely localize and delineate fine structures in images. In medical imaging, where capturing detailed information is crucial for accurate diagnosis and analysis, this limitation was particularly pronounced [91].

U-Net's innovative design integrates a contracting path, which captures context and abstract features, and an expansive path, which enables precise localization. The architecture incorporates skip connections that link corresponding layers in the contracting and expansive paths. These skip connections facilitate the integration of high-resolution features from the contracting path into the expansive path, aiding in the precise reconstruction of detailed structures. This unique structure earned U-Net its name due to its U-shaped architecture.

The skip connections can be represented as concatenating feature maps where

$$H_{\text{output}} = \text{concat}\left(H_{\text{contracting}}, H_{\text{expansive}}\right)$$

The contracting path involves standard convolutional and pooling operations, while the expansive path employs transposed convolutions for upsampling. Skip connections concatenate feature maps from the contracting path to the corresponding layers in the expansive path. The entire architecture is trained end-to-end with a loss function that compares the predicted segmentation mask to the ground truth.

U-Net's architecture, tailored with skip connections and adept utilization of convolutional and transposed convolutional layers, excels in semantic segmentation. Particularly beneficial in medical image analysis, U-Net finds applications in tasks demanding precise delineation of structures, such as organs or lesions. Its effectiveness extends to image segmentation, translation, and various domains requiring accurate object localization within images.

Author Contributions Jarrad Perron is listed as the corresponding author and so inquiries, questions, or corrections may be submitted directly to him. He was responsible for writing Sects. 1, 2, and 5. Iman Beheshti served as point of contact for initiating contributions to the current volume and was responsible for writing Sects. 3 and 4.

References

1. Eustance C (2003) Physics in nuclear medicine. In: Cherry SR, Sorenson JA, Phelps ME (eds) Saunders, 3rd edn, p 512. Hardbound, £52.99, ISBN 0–7216-8341-X," vol. 10, ed: Elsevier Ltd, 2004, pp 161–162

2. Rohren EM, Turkington TG, Coleman RE (2004) Clinical applications of PET in oncology. Radiology 231(2):305–332. https://doi.org/10.1148/radiol.2312021185

3. Salaün P-Y et al (2020) Good clinical practice recommendations for the use of PET/CT in oncology. Eur J Nucl Med Mol Imaging 47(1): 28–50. https://doi.org/10.1007/s00259-019-04553-8

4. Unterrainer M et al (2020) Recent advances of PET imaging in clinical radiation oncology.

Radiat Oncol (London, England) 15(1): 88–15. https://doi.org/10.1186/s13014-020-01519-1

5. Fonti R, Conson M, Del Vecchio S (2019) PET/CT in radiation oncology. Semin Oncol 46(3):202–209. https://doi.org/10.1053/j.seminoncol.2019.07.001

6. Chowdhury FU, Scarsbrook AF (2008) The role of hybrid SPECT-CT in oncology: current and emerging clinical applications. Clin Radiol 63(3):241–251. https://doi.org/10.1016/j.crad.2007.11.008

7. Hicks RJ, Hofman MS (2012) Is there still a role for SPECT-CT in oncology in the PET-CT era? Nat Rev Clin Oncol 9(12):

712–720. https://doi.org/10.1038/nrclinonc.2012.188

8. Perron J, Ko JH (2022) Review of quantitative methods for the detection of Alzheimer's disease with positron emission tomography. Appl Sci 12(22):11463. https://doi.org/10.3390/app122211463

9. Perovnik M et al (2022) Automated differential diagnosis of dementia syndromes using FDG PET and machine learning. Front Aging Neurosci 14:1005731. https://doi.org/10.3389/fnagi.2022.1005731

10. Perovnik M, Rus T, Schindlbeck KA, Eidelberg D (2023) Functional brain networks in the evaluation of patients with neurodegenerative disorders. Nat Rev Neurol 19(2):73–90. https://doi.org/10.1038/s41582-022-00753-3

11. Minoshima S, Mosci K, Cross D, Thientunyakit T (2021) Brain [F-18]FDG PET for clinical dementia workup: differential diagnosis of Alzheimer's disease and other types of dementing disorders. Semin Nucl Med 51(3):230–240. https://doi.org/10.1053/j.semnuclmed.2021.01.002

12. Katako A et al (2018) Machine learning identified an Alzheimer's disease-related FDG-PET pattern which is also expressed in Lewy body dementia and Parkinson's disease dementia. Sci Rep 8(1):13236–13213. https://doi.org/10.1038/s41598-018-31653-6

13. Beheshti I, Geddert N, Perron J, Gupta V, Albensi BC, Ko JH (2022) Monitoring Alzheimer's disease progression in mild cognitive impairment stage using machine learning-based FDG-PET classification methods. J Alzheimers Dis 89(4):1493–1502. https://doi.org/10.3233/JAD-220585

14. Spetsieris PG, Ma Y, Dhawan V, Eidelberg D (2009) Differential diagnosis of parkinsonian syndromes using PCA-based functional imaging features. NeuroImage (Orlando, Fla) 45(4):1241–1252. https://doi.org/10.1016/j.neuroimage.2008.12.063

15. Cummings J, Aisen P, Apostolova LG, Atri A, Salloway S, Weiner M (2021) Aducanumab: appropriate use recommendations. J Prev Alzheimers Dis 8(4):398–410. https://doi.org/10.14283/jpad.2021.41

16. Cummings J et al (2023) Lecanemab: Appropriate Use Recommendations. J Prev Alzheimers Dis 10(3):362–377. https://doi.org/10.14283/jpad.2023.30

17. Politis M (2014) Neuroimaging in Parkinson disease: from research setting to clinical practice. Nat Rev Neurol 10(12):708–722. https://doi.org/10.1038/nrneurol.2014.205

18. Hughes AJ, Daniel SE, Kilford L, Lees AJ (1992) Accuracy of clinical diagnosis of idiopathic Parkinson's disease: a clinicopathological study of 100 cases. J Neurol Neurosurg Psychiatry 55(3):181–184. https://doi.org/10.1136/jnnp.55.3.181

19. Booij J, Speelman JD, Horstink MWIM, Wolters EC (2001) The clinical benefit of imaging striatal dopamine transporters with [123I]FP-CIT SPET in differentiating patients with presynaptic parkinsonism from those with other forms of parkinsonism. Eur J Nucl Med 28(3):266–272. https://doi.org/10.1007/s002590000460

20. van Royen E, Verhoeff NFLG, Speelman JD, Wolters EC, Kuiper MA, Janssen AGM (1993) Multiple system atrophy and progressive Supranuclear palsy: diminished striatal D2 dopamine receptor activity demonstrated by 123I-IBZM single photon emission computed tomography. Archives Neurol (Chicago) 50(5):513–516. https://doi.org/10.1001/archneur.1993.00540050063017

21. Goffin KMD, Dedeurwaerdere SP, Van Laere KMDPD, Van Paesschen WMDP (2008) Neuronuclear assessment of patients with epilepsy. Semin Nucl Med 38(4):227–239. https://doi.org/10.1053/j.semnuclmed.2008.02.004

22. la Fougère C, Rominger A, Förster S, Geisler J, Bartenstein P (2009) PET and SPECT in epilepsy: a critical review. Epilepsy Behav 15(1):50–55. https://doi.org/10.1016/j.yebeh.2009.02.025

23. Donohoe KJ, Frey KA, Gerbaudo VH, Mariani G, Nagel JS, Shulkin B (1978) Procedure guideline for brain death scintigraphy. J Nucl Med 44(5):846–922. Author reply 922, 2004

24. MacDonald D, Stewart-Perrin B, Shankar JJS (2018) The role of neuroimaging in the determination of brain death. J Neuroimaging 28(4):374–379. https://doi.org/10.1111/jon.12516

25. Bermo M et al (2021) Utility of SPECT functional neuroimaging of pain. Front Psych 12:705242. https://doi.org/10.3389/fpsyt.2021.705242

26. Kung BT et al (2019) An update on the role of 18F-FDG-PET/CT in major infectious and inflammatory diseases. Am J Nucl Med Mol Imaging 9(6):255–273

27. Fendler WP, Czernin J, Herrmann K, Beyer T (2016) Variations in PET/MRI operations: results from an international survey among 39 active sites. J Nucl Med 57(12):2016–2021. https://doi.org/10.2967/jnumed.116.174169

28. Bailey DL et al (2016) Combined PET/MRI: from status quo to status go. Summary report of the fifth international workshop on PET/MR imaging; February 15–19, 2016; Tübingen, Germany. Mol Imaging Biol 18(5): 637–650. https://doi.org/10.1007/s11307-016-0993-2

29. Fraum TJMD, Fowler KJMD, McConathy JMDP (2016) PET/MRI. Acad Radiol 23(2): 220–236. https://doi.org/10.1016/j.acra.2015.09.008

30. Currie GM, Leon J, Nevo E, Kamvosoulis P (2022) PET/MRI, part 4: clinical applications. J Nucl Med Technol 50(2):90–96. https://doi.org/10.2967/jnmt.121.263288

31. States LJ, Reid JR (2020) Whole-body PET/MRI applications in pediatric oncology. Am J Roentgenol (1976) 215(3):713–725. https://doi.org/10.2214/AJR.19.22677

32. Institute of Medicine (US) Committee on Battlefield Radiation Exposure Criteria (2000) An evaluation of radiation exposure guidance for military operations: interim report, Compass Series, 1st edn. National Academies Press, Washington, D.C

33. Nofzinger E, Maquet P, Thorpy MJ (2013) Neuroimaging of Sleep and Sleep Disorders. Cambridge University Press

34. Limacher MC, Douglas PS, Germano G, Laskey WK, Lindsay BD, McKetty MH, Moore ME, Park JK, Prigent FM, Walsh MN (1998) Radiation safety in the practice of cardiology. J Am Coll Cardiol 31(4):892–913. https://doi.org/10.1016/S0735-1097(98)00047-3

35. Cherry SR, Dahlbom M (2006) PET: physics, instrumentation, and scanners, 1. Aufl. ed. Springer, New York

36. Jackson A (2003) Positron emission tomography. Basic science and clinical practice. By P E Valk, D L Bailey, D W Townscnd and M N Maisey, pp. xix+ 884. Springer-Verlag, London, UK., £115.00 ISBN 1–85233–485-1," vol. 77, ed: British Institute of Radiology, 2004, pp 704–704

37. Rahmim A, Zaidi H (2008) PET versus SPECT: strengths, limitations and challenges. Nucl Med Commun 29(3):193–207. https://doi.org/10.1097/MNM.0b013e3282f3a515

38. Adak S, Bhalla R, Vijaya Raj KK, Mandal S, Pickett R, Luthra SK (2012) Radiotracers for SPECT imaging: current scenario and future prospects. Radiochim Acta 100(2):95–107. https://doi.org/10.1524/ract.2011.1891

39. Maddahi J (2012) Properties of an ideal PET perfusion tracer: new PET tracer cases and data. J Nucl Cardiol 19(Suppl 1):30–37.

40. Schwaiger M, Wester H-J (2011) How many PET tracers do we need? J Nucl Med 52(Supplement 2):36S–41S. https://doi.org/10.2967/jnumed.110.085738

41. Portnow LH, Vaillancourt DE, Okun MS (2013) The history of cerebral PET scanning: from physiology to cutting-edge technology. Neurology 80(10):952–956. https://doi.org/10.1212/WNL.0b013e318285c135

42. Pimlott SL, Sutherland A (2011) Molecular tracers for the PET and SPECT imaging of disease. Chem Soc Rev 4(1):149–162. https://doi.org/10.1039/b922628c

43. Mosconi L et al (2009) FDG-PET changes in brain glucose metabolism from normal cognition to pathologically verified Alzheimer's disease. Eur J Nucl Med Mol Imaging 36(5): 811–822. https://doi.org/10.1007/s00259-008-1039-z

44. Kato T, Inui Y, Nakamura A, Ito K (2016) Brain fluorodeoxyglucose (FDG) PET in dementia. Ageing Res Rev 30:73–84. https://doi.org/10.1016/j.arr.2016.02.003

45. Nestor PJ et al (2018) Clinical utility of FDG-PET for the differential diagnosis among the main forms of dementia. Eur J Nucl Med Mol Imaging 45(9):1509–1525. https://doi.org/10.1007/s00259-018-4035-y

46. Woyk K et al (2023) Brain 18F-FDG-PET and an optimized cingulate Island ratio to differentiate Lewy body dementia and Alzheimer's disease. J Neuroimaging 33(2):256–268. https://doi.org/10.1111/jon.13068

47. Dave A, Hansen N, Downey R, Johnson C (2020) FDG-PET imaging of dementia and neurodegenerative disease. Semin Ultrasound CT MRI 41(6):562–571. https://doi.org/10.1053/j.sult.2020.08.010

48. Jack CR et al (2018) NIA-AA research framework: toward a biological definition of Alzheimer's disease. Alzheimers Dement 14(4): 535–562. https://doi.org/10.1016/j.jalz.2018.02.018

49. Oldan JD, Jewells VL, Pieper B, Wong TZ (2021) Complete evaluation of dementia: PET and MRI correlation and diagnosis for the Neuroradiologist. AJNR Am J Neuroradiol 42(6):998–1007. https://doi.org/10.3174/ajnr.A7079

50. van Waarde A, Marcolini S, de Deyn PP, Dierckx RAJO (2021) PET agents in dementia: an overview. Semin Nucl Med 51(3):196–229. https://doi.org/10.1053/j.semnuclmed.2020.12.008

51. Okamura N, Harada R, Ishiki A, Kikuchi A, Nakamura T, Kudo Y (2018) The development and validation of tau PET tracers: current status and future directions. Clin Trans Imag Rev Nuclear Med Mol Imag 6(4):305–316. https://doi.org/10.1007/s40336-018-0290-y

52. Ossenkoppele R et al (2021) Accuracy of tau positron emission tomography as a prognostic marker in preclinical and prodromal Alzheimer disease: a head-to-head comparison against amyloid positron emission tomography and magnetic resonance imaging. JAMA Neurol 78(8):961–971. https://doi.org/10.1001/jamaneurol.2021.1858

53. I. Beheshti *et al.*, "Accurate lateralization and classification of MRI-negative 18F-FDG-PET-positive temporal lobe epilepsy using double inversion recovery and machine-learning" (in eng), Comput Biol Med, vol. 137, p. 104805, 10 2021, doi:https://doi.org/10.1016/j.compbiomed.2021.104805

54. I. Beheshti, S. Nugent, O. Potvin, and S. Duchesne, "Disappearing metabolic youthfulness in the cognitively impaired female brain" (in eng), Neurobiol Aging, vol. 101, pp. 224–229, 05 2021, doi:https://doi.org/10.1016/j.neurobiolaging.2021.01.026

55. Lau A, Beheshti I, Modirrousta M, Kolesar TA, Goertzen AL, Ko JH (2021) Alzheimer's disease-related metabolic pattern in diverse forms of neurodegenerative diseases. Diagnostics 11(11):2023

56. Devous MD Sr (1995) SPECT functional brain imaging; technical considerations. J Neuroimaging 5(s1):s2–s13

57. Warwick JM (2004) Imaging of brain function using SPECT. Metab Brain Dis 19:113–123

58. Henderson TA, Cohen PF, Cardaci G, Urbain J-LC (2022) The emerging role of SPECT functional neuroimaging in Psychiatry & Neurology. Front Psych 13:928653

59. Ferrando R, Damian A (2021) Brain SPECT as a biomarker of neurodegeneration in dementia in the era of molecular imaging: still a valid option? Front Neurol 12:629442

60. Jiang J et al (2018) Study of the influence of age in 18 F-FDG pet images using a data-driven approach and its evaluation in Alzheimer's disease. Contrast Media Mol Imaging 2018

61. Cecchin D, Garibotto V, Law I, Goffin K (2021) PET imaging in neurodegeneration and neuro-oncology: variants and pitfalls. Semin Nucl Med 51(5) Elsevier:408–418

62. Minoshima S, Cross D, Thientunyakit T, Foster NL, Drzezga A (2022) 18F-FDG PET imaging in neurodegenerative dementing disorders: insights into subtype classification, emerging disease categories, and mixed dementia with copathologies. J Nucl Med 63 (Supplement 1):2S–12S

63. Michopoulou SK et al (2023) Brain PET and SPECT imaging and quantification: a survey of the current status in the UK. Nucl Med Commun 44(10):834–842

64. Beheshti I et al (2020) Pattern analysis of glucose metabolic brain data for lateralization of MRI-negative temporal lobe epilepsy. Epilepsy Res 167:106474. https://doi.org/10.1016/j.eplepsyres.2020.106474

65. Sone D et al (2021) "neuroimaging-based brain-age prediction in diverse forms of epilepsy: a signature of psychosis and beyond" (in eng). Mol Psychiatry 26(3):825–834. https://doi.org/10.1038/s41380-019-0446-9

66. LeCun Y, Bengio Y, Hinton G (2015) Deep learning. Nature (London) 521(7553): 436–444. https://doi.org/10.1038/nature14539

67. Heaton J (2018) Ian Goodfellow, Yoshua Bengio, and Aaron Courville: deep learning: the MIT press, 2016, 800 pp, ISBN: 0262035618. Genet Program Evolvable Mach 19(1–2):305–307. https://doi.org/10.1007/s10710-017-9314-z

68. Yao Y, Yu B, Gong C, Liu T (2023) Understanding how Pretraining regularizes deep learning algorithms. IEEE Trans Neural Networks Learn Syst 34(9):5828–5840. https://doi.org/10.1109/TNNLS.2021.3131377

69. Baker B, Gupta O, Naik N, Raskar R (2016) Designing neural network architectures using reinforcement learning. arXivorg. https://doi.org/10.48550/arxiv.1611.02167

70. Hwang G et al (2020) "Brain aging in temporal lobe epilepsy: chronological, structural, and functional" (in eng). NeuroImage Clinical 25: 102183. https://doi.org/10.1016/j.nicl.2020.102183

71. Nenning K-H, Langs G (2022) Machine learning in neuroimaging: from research to clinical practice. Die Radiol 62(Suppl 1):1–10

72. Davatzikos C (2019) Machine learning in neuroimaging: Progress and challenges. NeuroImage 197:652–656

73. Apicella A, Donnarumma F, Isgrò F, Prevete R (2021) A survey on modern trainable activation functions. Neural Netw 138:14–32. https://doi.org/10.1016/j.neunet.2021.01.026

74. Janocha K, Czarnecki WM (2017) On loss functions for deep neural networks in

classification. arXiv.org. https://doi.org/10.48550/arxiv.1702.05659

75. Yamashita R, Nishio M, Do RKG, Togashi K (2018) Convolutional neural networks: an overview and application in radiology. Insights Imaging 9(4):611–629. https://doi.org/10.1007/s13244-018-0639-9

76. Ruder S (2016) An overview of gradient descent optimization algorithms. arXiv.org. https://doi.org/10.48550/arxiv.1609.04747

77. Nesterov Y (2009) Primal-dual subgradient methods for convex problems. Math Program 120(1):221–259. https://doi.org/10.1007/s10107-007-0149-x

78. Ioannou G, Tagaris T, Stafylopatis A (2023) AdaLip: an adaptive learning rate method per layer for stochastic optimization. Neural Process Lett 55(5):6311–6338. https://doi.org/10.1007/s11063-022-11140-w

79. Kingma DP, Ba J (2014) Adam: a method for stochastic optimization. arXiv.org. https://doi.org/10.48550/arxiv.1412.6980

80. Kukačka J, Golkov V, Cremers D (2017) Regularization for deep learning: a taxonomy. arXiv.org. https://doi.org/10.48550/arxiv.1710.10686

81. Hinton GE, Srivastava N, Krizhevsky A, Sutskever I, Salakhutdinov RR (2012) Improving neural networks by preventing co-adaptation of feature detectors. arXiv.org. https://doi.org/10.48550/arxiv.1207.0580

82. Berrar D (2019) Cross-validation. Elsevier Inc, pp 542–545

83. Leibe B, Matas J, Sebe N, Welling M (2016) Identity mappings in deep residual networks, Lecture Notes in Computer Science, vol 9908. Springer, Switzerland, pp 630–645

84. Huang G, Liu Z, van der Laurens M, Weinberger KQ (2016) Densely connected convolutional networks. Ithaca. Cornell University Library, arXiv.org. https://doi.org/10.48550/arxiv.1608.06993

85. Hinton GE, Salakhutdinov RR (2006) Reducing the dimensionality of data with neural networks. Science (Am Assoc Adv Sci) 313(5786):504–507. https://doi.org/10.1126/science.1127647

86. Doersch C (2016) Tutorial on Variational autoencoders. arXiv.org. https://doi.org/10.48550/arxiv.1606.05908

87. Goodfellow I et al (2020) Generative adversarial networks. Commun ACM 63(11):139–144. https://doi.org/10.1145/3422622

88. Mirza M, Osindero S (2014) Conditional generative adversarial nets. arXiv.org. https://doi.org/10.48550/arxiv.1411.1784

89. Salehinejad H, Sharan S, Barfett J, Colak E, Valaee S (2017) Recent advances in recurrent neural networks. arXiv.org. https://doi.org/10.48550/arxiv.1801.01078

90. Cheng J, Li D, Lapata M (2016) Long short-term memory-networks for machine Reading. arXiv.org. https://doi.org/10.48550/arxiv.1601.06733

91. Ronneberger O, Fischer P, Brox T (2015) U-net: convolutional networks for biomedical image segmentation. arXiv.org. https://doi.org/10.48550/arxiv.1505.04597

INDEX

Daichi Sone (ed.), *Molecular Imaging for Brain Diseases*, Neuromethods, vol. 222, https://doi.org/10.1007/978-1-0716-4494-2,
© The Editor(s) (if applicable) and The Author(s), under exclusive license to Springer Science+Business Media, LLC, part of Springer
Nature 2025